TECHNOSCIENTIFIC ANGST

TECHNOSCIENTIFIC ANGST

ethics + responsibility

Raphael Sassower

 University of Minnesota Press
Minneapolis
London

Published by the University of Minnesota Press
111 Third Avenue South, Suite 290
Minneapolis, MN 55401-2520
http://www.upress.umn.edu

Printed in the United States of America on acid-free paper

Library of Congress Cataloging-in-Publication Data

Sassower, Raphael.
    Technoscientific angst : ethics and responsibility / Raphael Sassower.
        p.   cm.
    Includes bibliographical references and index.
    ISBN 0-8166-2956-0 (hc : alk. paper). — ISBN 0-8166-2957-9 (pb : alk. paper)
    1. Technology—Philosophy. 2. Technology—Moral and ethical aspects.
    3. Science—Philosophy. 4. Science—Moral and ethical aspects. I. Title.
    T14.S268   1997
    174′.96—dc21   97-20529
                                                                                   CIP

Dedicated to the memory of
Donald Campbell and Ernest Gellner

Contents

Preface

The anguish of artists and poets is celebrated by societies that expect justice and happiness in the future regardless of their current conditions. Anguish is accepted and endorsed not so much as a judgment about the present but as a means to envision and usher in a different future. Oddly enough, those who are members of the technoscientific community are discouraged from playing the same social role as do artists and poets; their anguish is neither acknowledged nor displayed. On the rare occasions when they express professional anxiety, personal anguish, or cultural angst, they are invited to leave the technoscientific community. I find this situation unfortunate, disturbing, and socially harmful. It is reasonable to believe that if members of the technoscientific community were encouraged to display their concerns publicly and thereby enhance the critical involvement of society as a whole (as did, for example, Joseph Rotblat, the 1995 Nobel Peace Prize laureate), we might be spared in the future the horrors of the past, like those of Auschwitz and Hiroshima.

How can we expect to avert future horrors from taking place? In *Cultural Collisions: Postmodern Technoscience* (1995), I used the metaphor of the European café to explain the multitudes of language games and cultural rules of etiquette that govern our behavior in different geographic locations. I argued for the need for philosophers or philosophically minded intellectuals to translate from one language to another so as to minimize, if not completely eliminate, the potential for misunderstanding that leads to hatred and death. In the present context, I fear the metaphor of a café will be of little solace to those who suffered in concentration camps or at Hiroshima or Nagasaki, or to any of their children. Perhaps a more forceful metaphor with a preemptive thrust is necessary to highlight the urgency of the current situation of the technoscientific community.

The airplane (as a metaphor and as a lived experience) embodies many of the issues under discussion in this book. To begin with, it captures images of Icarus, who glued feathers to his arms with wax and soared to the sky, only to find out that as he got closer to the sun, the wax melted and destroyed his artificially constructed wings. The image is captivating, the dream comes true, but the technology falls short of the conceptual scheme. Second, once the airplane was designed and developed, its eventual uses remained beyond the control of technoscientists: Would it fly people or bombs? Would it save or destroy life? Third, it is not clear whether air travel has enhanced or retarded world peace: Is the so-called global village a better place to live in than the world of an earlier century wherein travel was difficult and infrequent? And fourth, one can ask whether airplanes have contributed to the democratization of the world. Though air travel is common, it is still stratified into travel by luxury and low-fare airlines or into travel in first class, connoisseur class, business class, or economy class. Similarly, though wealthy countries have military might displayed in aerial superiority (from air force planes to satellites), even the poorest of the nation-states have aircraft and weaponry capable of destroying millions of lives.

Sitting some 35,000 feet above ground, watching the clouds and the sky, one has the opportunity to examine the old-fashioned views of the conquest of nature. Gravity seems to dissipate the higher the plane flies, and vast distances are covered in a matter of minutes and hours. Traditional cultural matrices of time and space are reconfigured by the airplane. A technoscientific feat has become an inescapable reality, a reality whose control is no longer in the hands of particular organizations and individuals. This does not mean that aerial and space developments are beyond human control; all it means is that the clock cannot be set back to a time when only birds flew over the horizon, that it is now impossible to imagine a world without airplanes. Just as gunpowder, once invented, never left us (we only went on to more sophisticated modes of explosion and destruction), so the airplane is here to stay. We may have planes perform the task of buses and trains, we may build them bigger or smaller, more or less safe, quicker or slower, but their very presence remains intact.

This all connects to what I am trying to explore in the present text. One need not be either a technophile or a technophobe to appreciate the cultural revolution brought about in the twentieth century with the advent of the airplane. Instead, one may consider the simple experience of flying on an airplane. Why choose this mode of transportation? Have other alternatives been considered? How do fellow travelers interact? Do they talk to each other? Are they kind to each other? Do they ever consider the fact that their conversations with the persons alongside them may be the last they will ever have with anyone? This is different from the chance conversation one has with people at the café because the cessation of that conversation is voluntary under most circumstances (earthquakes and other unforeseen disasters notwithstanding). By contrast, one usually has no choice where to sit in an airplane, so one's neighbor is arbitrarily chosen (some airlines have changed this policy); likewise, the passenger has no choice of how long to stay seated and cannot leave at will; and finally, the passenger has no control over the fate of the flight, from weather turbulence to crashes. One's behavior on an airplane, therefore, is a useful image with which to examine our behavior in general, that is, our quest for control over an ordered universe.

It is a useful image because it shows us the extent to which we can claim responsibility for our own actions and ascribe responsibility to others. It is a metaphor for our helplessness amid the control we try to exert over our surroundings. It is a reality both delightful and frightening, but a reality that we cannot imagine being without. In short, an airplane ride is inherently and more pronouncedly unpredictable than one would wish to believe. The cultural maturation of the past few decades (post–World War II) has allowed us to face this techno-scientific achievement with responsibility couched in social and economic terms. For example, instead of being in awe of the flying carpets or mechanized birds, we are asking ourselves how much money we are willing to spend to make air travel relatively risk free or how much money we are willing to spend to have our skies clear of nuclear threats.

When we answer in particular ways and compare the accident rates of cars and airplanes or the probability of a nuclear holocaust, we are already in the midst of a negotiated compromise on the future of public transportation and national security. As long as the compromise is reached openly, publicly, and rationally, it is likely to encourage continued criticism and pressure for future revision and improvement. Otherwise, every action (whether resulting in disaster or not) will be prone to public scorn and the obfuscation of professional and personal responsibility.

This book is motivated by and concerned with the responsibility that intellectuals share in regards to the development of science and technology at the end of this century. The horrors of World War II went beyond those of other wars primarily because of the devices that were used: the atomic bomb in the cases of Hiroshima and Nagasaki and the gas chambers in Nazi concentration camps. This book is neither devoted to recounting these horrors (because they are well documented elsewhere) nor concerned with assigning personal and collective guilt.

Instead, I wish to examine in this book two related topics: first, the inherent scientific anguish and angst displayed by scientists and suppressed by the public in relation to technological developments over which scientists have indirect control; and second, the ambiguity concerning the responsibility of members of the technoscientific community in the face of mass destruction. These topics are related because of broad cultural expectations about control of an ordered universe and the inherent ambiguity of scientific knowledge claims. Moreover, the image of science as the citadel of truth, objectivity, and value neutrality—that is, what has come to be known as the scientific ethos of self-criticism and self-policing—is simultaneously being upheld and undermined. It is important to go beyond or transform the image of science so that a critique of technoscience will address critical issues, such as nuclear disarmament and environmental stability, in ways that would enhance public debate and participation.

The forceful logic of the modern ideals of freedom and equality inadvertently leads to the stupefying reality of devastation. To achieve these ideals, institutions were formed that eventually precluded the interests of one class over another (to respond to Karl Marx) and adopted the faceless demeanor of an efficient bureaucracy (to respond to Max Weber). The classless and faceless order of the modern nation-state could simultaneously accommodate the fascist tendencies of the rigor and rigidity of modernity (with its dehumanizing calculus) as well as the emancipatory tendencies of modernity's liberalism.

Political leaders are elected to their positions of power in order to fulfill the ideological mission and promise of the modern nation-state. They tend to accept the judgments of technoscientists because they expect them in turn to construct and deliver expedient strategies that will solve public problems. In some uncanny manner, Jews and Japanese were reduced during World War II, by fascist and democratic governments alike, to social variables that threatened the image of a perfect world order; as such, they had to be "dealt with" (excluded, targeted, and annihilated). While Nazi Germany relied on trains and gas chambers, the United States relied on airplanes and atomic bombs. Both governments enlisted the services of qualified technoscientists. The means were not the same, of course, nor were the results. So I am not claiming moral equivalence between the cases. Yet the logic of destruction, the utilitarian calculus gone mad, inflicted pain and suffering similarly. Systematic victimization resulted in both cases in the injustice of premature death.

The critiques of science with which I am familiar (e.g., Marxist, feminist, postmodernist, social, and political critiques) tend to "throw the baby out with the bathwater" in the sense that they no longer search for ways to induce the scientific community to help police itself on ethical grounds. The standard argument is set up in the following manner: Science cannot watch over itself because it has an interest in its own proliferation; therefore, society cannot trust science and must police it from the outside. But what if society itself cannot be trusted, as in the case of Nazi Germany? What remains is a set of ideologically biased lobbying groups whose power to determine public policy is self-serving. Is there no way to revisit the ideals of science, whose intent was to create a more reasonable and open culture free from prejudices, when it decides on scientific policies?

As I demonstrated in *Cultural Collisions*, there are crucial political and economic issues to be considered when deciding on public projects, such as the Superconducting Supercollider. The significance of the issues I try to raise here, however much they are inspired by the horrors of World War II, go beyond philosophical circles and will be of interest to a broader group of thinkers and policy makers. So I do not wish to merely add my concerns to those raised at the fiftieth-anniversary commemoration of the dropping of the atomic bomb and the liberation of Auschwitz. Instead, I wish to use philosophical insights and argumentative techniques of critique to frame the parameters within which the scientific community can become publicly responsible for its cultural contributions. In short, I offer my translation of multiple discourses and language games into one text. I use diaries and archival material, scholarly texts and plays performed in theaters, popular and journalistic stories, and narratives of all sorts.

The book is divided into seven chapters, but the order in which they are published does not limit the options of the order of reading them. Some may choose to follow the order of presentation, which moves from the historical and ideological antecedents to World War II through an explanation of the cultural expectations that informed our cultural development, right into a discussion of the rationale for mass destruction in the German and American contexts. Others may choose to shift their attention after the first two chapters to the fifth chapter, wherein a reconstruction of the modern–postmodern situation is offered. Still others may choose to skip as soon as possible to the sixth chapter and ascertain the conditions under which ethical norms may be established in the postmodern and atomic age. In general, what these chapters offer is an explanatory model for the events labeled "Auschwitz" and "Hiroshima," a model that gives rise to practical recommendations for future behavior.

I interlace my critique with deep empathy for members of the technoscientific community, without thereby relinquishing the demand that they assume personal responsibility for the fruits of technoscience. When doing so, I move away from the endorsement of postmodernism as a relativist escape and toward an incorporation of the postmodern ethos under conditions of ethical commitment in the face of the ambiguities underlining the human condition. I view these chapters, then, as contributing elements in the construction of a workable platform or agenda for a culture surrounded by technoscientific angst.

I must thank the Committee on Research and Creative Works at the University of Colorado at Colorado Springs for its generous support which allowed me to complete my research in the summer of 1996. I would also like to thank David Hawkins for his helpful and critical comments and for sharing personal memories from his days at Los

Alamos. Among those who read earlier drafts, I wish to thank Mary Ann Cutter, Robert Proctor, Ray Spier, Todd Macon, Leah Stepanek, and the anonymous reviewers of the University of Minnesota Press. My seminar students have given me useful advice about what matters, what makes provocative reading, and what is worth fighting for. I thank them all, and take full responsibility for my omissions and mistakes.

Special thanks go to Louis Cicotello, who designed the collage for the cover of the book. His quiet and critical support of and collaboration with my work in words and images are heartfelt and greatly appreciated.

I am sorry that I cannot thank the librarians in charge of the archives at Los Alamos. Their dismissal of my repeated requests (in person and in writing) under the pretense that "there isn't much of interest here" and that "most of what we have here is still classified, and whatever is declassified is already available in print" seemed unreasonable. Prior to this experience, I never understood how sensitive our military and political institutions remain when so-called national security issues are at stake. It seems that the lessons of the Manhattan Project and the pain associated with them remain unclear or ignored. I hope my own effort will contribute to a more accessible power structure.

Parts of Chapter 1 were published earlier as "Responsible Technoscience: The Haunting Reality of Auschwitz and Hiroshima," *Science and Engineering Ethics* 2(3) (1996): 277–290. I thank the editors for permission to reprint these sections.

Responsible Technoscience: The Haunting Reality of Auschwitz and Hiroshima

Background

The commemoration in 1995 of the fiftieth anniversaries of two major events of World War II, the liberation of Auschwitz (January 1945) and the dropping of the atomic bomb on Hiroshima (August 1945), invites us to use these events as starting points for self-examination. I focus here only on the "technoscientific" facets of these two events, and not on the entire historical background of the rise of fascism, the period between the two world wars, and the circumstances surrounding World War II (see, e.g., Dawidowicz 1975, Gilbert 1985, and Hilberg 1973). Furthermore, my concern is not with the relationship between science, technology, and politics in terms of power structures and capitalist industrialism but, rather, with the self-policing functions of the technoscientific community. Finally, as the following four brief outlines illustrate, though laudable intentions may be ascribed to this community and though we may applaud its cultural ascent, it remains painfully inept when assuming responsibility for horrible events such as the use of gas chambers in Auschwitz and the nuclear bombing of Hiroshima and Nagasaki. Norbert Wiener reminds us that

> the Greeks regarded the act of discovering fire with very split emotions.
> On the one hand, fire was for them as for us a great benefit to all hu-
> manity. On the other, the carrying down of fire from heaven to earth
> was a defiance of the Gods of Olympus, and could not but be punished

by them as a piece of insolence towards their prerogatives . . . The sense of tragedy is that the world is not a pleasant little nest made for our protection . . . It is a dangerous world, in which there is no security . . . If a man with this tragic sense approaches, not fire, but another manifestation of original power, like the splitting of the atom, he will do so with fear and trembling. He will not leap in where angels fear to tread, unless he is prepared to accept the punishment of the fallen angels. Neither will he calmly transfer to the machine made in his own image the responsibility for his choice of good and evil, without continuing to accept full responsibility for that choice. (Wiener 1989, 184)

Perhaps we have forgotten the sense of tragedy that permeated the cultures of ancient times; perhaps we have become accustomed to overlooking the sadness within our joyful implementation of new technologies; perhaps we have forgotten the limits of our humanity. Moreover, when we do appreciate our limits and worry about the consequences of our productions, we seem to relegate our worries to others, pretending that the potential for the military use of techno-scientific inventions justifies critical silence in the face of national security. As Wiener comments:

It is the great public which is demanding the utmost of secrecy for modern science in all things which may touch its military uses. This demand for secrecy is scarcely more than the wish of a sick civilization not to learn of the progress of its own disease. So long as we can continue to pretend that all is right with the world, we plug up our ears against the sound of "Ancestral voices prophesying war." (Wiener 1989, 127)

Technoscience

Science, technology, and engineering are not separate activities undertaken in separate communities; instead, they influence and enhance each other's development in fundamental ways. For example, technical instruments are crucial for theoretical breakthroughs, and a conceptual background is essential for engineering applications. It therefore makes sense to speak of their constellation as "technoscience," as Jean-François Lyotard does:

In the present epoch, science and technology combine to form contemporary technoscience. In technoscience, technology plays the role of furnishing the proof of scientific arguments: it allows one to say of a scientific utterance that claims to be true, "here is a case of it." The result of this is a profound transformation in the nature of knowledge. Truth is subjected to more and more sophisticated means of "falsifying" scientific utterances. (Lyotard 1982, 14–15)

The combination of the terms "science" and "technology" into "technoscience" is still explained by Lyotard in traditional terms—technology is the implementation of science, a form of testing scientific knowledge—and not as a blurred site of knowledge production, that is, not as a site wherein one cannot do the one without the other, as Bruno Latour and Steve Woolgar describe it (1986). Technoscience, then, denotes a dynamic relationship among instruments and people within a cultural context that brings about conceptual and practical changes.

Technoscience, as the constellation of science, technology, and engineering, is best understood in terms of the activities of members of a community. Whether one follows the descriptions of Robert Merton (1973), Michael Polanyi (1966), or Thomas Kuhn (1970), an appreciation of the activities of members of the technoscientific community helps society demystify technoscience. As we shall see in later chapters, the Manhattan Project was an excellent example of scientists being technicians and engineers, especially once Los Alamos became the center of activity. Since then, we can more clearly talk about the activities of the technoscientific community, as opposed to technoscience as such, because we realize that we are dealing with humans and their creations under specific sets of material conditions. Reminding ourselves that there is a human face attached to technoscience helps us recognize the inherent fallibility of technoscience and the need for ongoing critical evaluations and revisions. In short, the assignment of responsibility may shift depending on particular circumstances, but this does not mean that it is ever justified to fail to assign responsibility.

The technoscientific community is engaged in research and development activities whose scope may elude individual members, especially in "Big Science" projects, like the Manhattan Project. The scope of a project may be so vast that it raises two related problems: (1) Those involved in research may not be those involved in development, and (2) within either research or development, no individual member's contribution is fully appreciated and comprehended by every other member. Though the general contours of the project may be outlined for all participants at different stages, the details of each stage are known primarily to those directly involved in that stage.

The work of the technoscientific community—from the inception of ideas (whether out of curiosity or because of particular incentives, such as to win a war, patent a discovery, eliminate a disease, or make money) to the implementation of these ideas by industry or government—can be characterized as a seamless web of connections. Just as individual members of the community are not in a position to have a precise overview of the entire community, so they are not in a position to clearly divide any technoscientific process into discrete parts.

That it involves a community too large and heterogeneous to monitor and a process too fluid to control may lead one to think of technoscience as being autonomous in Langdon Winner's sense:

> Developments in the technical sphere continually outpace the capacity of individuals and social systems to adapt . . . One symptom of [these developments is] the belief that somehow technology has gotten out of control and follows its own course, independent of human direction. (Winner 1977, 3–13)

However accurate or inaccurate, dramatic, or provocative this description of the technoscientific community may be in the modern age of postindustrial, efficient productivity, it should not absolve the members of the technoscientific community from responsibility for their actions. Instead, as I will argue throughout this book, it should heighten their sense of personal responsibility.

Autonomy and the "Invisible Hand" of Modernity

One of the appealing aspects of the Enlightenment project (in its narrow and homogenizing definition) was its promise to overlook differences (in class affiliation, in origins, in religious beliefs, in physical appearance, and in mental aptitude) in the name of the ideals of equality and liberty. But in order to achieve these ideals, whether through education or assimilation, certain variables have to be erased, or at least ignored and overlooked. For example, personal traits and backgrounds are better left outside the educational apparatus if we are to treat everyone alike and provide equal opportunities to everyone, say in Immanuel Kant's sense (1970).

The modern nation-state is a political apparatus that was established in part to legally guarantee equality and liberty: Ethnic and religious differences were to be handled in the private sphere so that the public sphere would be able to accommodate everyone equally and without interfering with anyone's liberty, as G. W. F. Hegel (1967) and John Stuart Mill (1984) argued respectively. (Whether or not these ideals have been properly applied over the years remains an open question.) The drive for the separation of state and church could guarantee freedom to all (as, for example, the U.S. Constitution of 1787 and the French Revolution of 1789 intended). And finally, politics and education were linked to the social and economic realms, for the capitalist system promised equal opportunity to anyone willing to compete and work hard; thus feudal and traditionalist obstacles could be overcome.

Adam Smith introduced the notion of the division of labor and the famous "invisible hand" as essential components for the free exchange of goods and services in the marketplace (Smith 1937, 423). The capitalist ideals of anonymity and autonomy (in whose name neoclassical political economists revisited Smith's ideals in the early twentieth century) were subverted into the reality of a faceless bureaucracy. Whereas Smith endorsed competition and free trade, Max Weber, a century later, recognized the institutional setting that ensures productivity and efficiency in the form of public and private bureaucracies (Weber 1968). Whether Smith's ideals are instantiated in a system of autonomous entrepreneurs whose individuality inadvertently benefits society or in a dehumanized bureaucracy that deals fairly with all, there remains the specter of an "invisible hand" to which no particular accountability or responsibility can be attached.

Modernity, with its Weberian instrumental rationality, its bureaucratic aura, and its cult of efficiency in Kolakowski's sense (1990), simply extended its tenets and commitments during the Holocaust:

> The "Final Solution" did not clash at any stage with the rational pursuit of efficient, optimal goal-implementation. On the contrary, *it arose out of a genuinely rational concern, and it was generated by bureaucracy true to its form and purpose.* (Bauman 1991a, 17)

Concentration camps are not to be seen as an aberration in any sense of the term: They are extremely disturbing but absolutely rational manifestations of the concern with the Jewish problem. As the Nuremberg trials at the end of World War II illustrated, a bureaucratic "invisible hand" was at work to exterminate millions of innocent citizens (Annas and Grodin 1992). Hannah Arendt (1963) aptly wondered in her account of the Eichmann trial in Jerusalem whether or not it was not possible to ascribe responsibility for the "banality of evil."

The Rationality of the Technoscientific Community

While the proponents of the Enlightenment promised the ideals of equality, liberty, and perpetual peace, later disciples had to accept equal opportunity, relative liberty, and roundtables for speech communities (in Jürgen Habermas's sense, 1979). A similar narrative can be reconstructed for the ideals of science and later developments in the technoscientific community. It is not that the ideals, such as objectivity, value neutrality, and truth, have vanished from this community; rather, what has become apparent is the price that may be paid in the name of these ideals. A negotiated compromise has been achieved over the years, at times under public pressure and with full disclosure and at other times clandestinely.

According to Robert Merton, the so-called father of American sociology of science, the "ethos of science" includes a complex system of values and norms that are "expressed in the form of prescriptions, proscriptions, preferences, and permissions," all which are "legitimized in terms of institutional values" (Merton 1973, 269). The particular norms of science—whether construed as heuristic ideals or as achievable goals—are classified into four categories: universalism, "communism," disinterestedness, and organized skepticism. These four categories may be revised over time and their labels may be changed, yet they capture the main thrust of the orientation of the technoscientific community.

It is essential to pause and review this system of values and norms for three reasons. First, it is supposed to inform and direct the initiating process undertaken by all aspiring members of this community. Second, it is a useful measure or set of criteria according to which the leadership of this community can monitor and self-police all other members and subgroups. And third, the interaction between this community and the public at large is based on and measured according to these categories. So Merton's outline becomes essential for those working within this community, those interacting with it, and those who wish to evaluate the workings of this community (both self-policing and external policing).

According to Merton, "universalism" means that "the acceptance or rejection of claims entering the lists of science is not to depend on the personal or social attributes of their protagonist . . . Objectivity precludes particularism." "Communism" means that "the substantive findings of science are a product of social collaboration and are assigned to the community. They constitute a common heritage." "Disinterestedness" means for Merton an "unusual degree of moral integrity," an "ultimate accountability of scientists to their compeers," because "the activities of scientists are subject to rigorous policing." Finally, "organized skepticism" is "both a methodological and an institutional mandate," and it manifests itself as "the temporary suspension of judgment and the detached scrutiny of beliefs in terms of empirical and logical criteria"(Merton 1973, 270–278).

Merton's categories reflect a strong commitment by members of the technoscientific community, which has a tradition of its own. The tradition was and still is set up to be distinguished from those of other communities, such as churches or commercial guilds. Instead of appealing to an indisputable divine authority (as do churches) or appealing to the self-interest of a fraternity of skilled laborers (as do guilds), the technoscientific community appealed to the truth and open criticism. The truth may sanction a model's authority but only provisionally; a counterexample or a new model may displace one truth with another. Though at times as powerful as churches and guilds, the technoscientific community, according to this description of its ethos, continues to invite challenges to its ideas and authority. Its power rests on its acceptance of the inherent ambiguity underlying all knowledge claims and the limitations of the human intellect.

Though the ethos of science, as Merton describes it, includes features that reflect the Enlightenment ideals in terms of the equality among scientists and their freedom to join the community and to critically evaluate its products, it also ensures a certain level of personal detachment (despite the call for high moral integrity). That is, trying to depersonalize scientific activities may entail unforeseen consequences. As far as Zygmunt Bauman is concerned, the modern interpretation and practice of the ideals of science are as follows:

> We need to take stock of the evidence that *the civilizing process is, among other things, a process of divesting the use and deployment of violence from moral calculus, and of emancipating the desiderata of rationality from interference of ethical norms or moral inhibitions.* (Bauman 1991a, 28)

If rationality is separated from the emotions and if instrumental rationality is separated from morality as such, then it makes perfect sense to speak of the separation between science and ethics. This separation, incidentally, differs from Merton's contention that "the scientific investigator does not preserve the cleavage between the sacred and the profane, between that which requires uncritical respect and that which can be objectively analyzed" (Merton 1973, 277–278). Though the scientific investigator finds it appropriate to investigate anything and not to shield any view from criticism, this still does not mean that there is an overlap (methodologically or institutionally) between empirical and ethical concerns. (The failure to recognize the bifurcation between the "is" and the "ought," as G. E. Moore reminds us, leads many to commit the Naturalistic Fallacy; see Sellars and Hospers 1970.) Though the intention of the leaders of the Enlightenment and their disciples over the years may have been honorable and though the ethos of science seems to account for "moral integrity," as Merton calls it, Bauman reminds us that "the self-imposed moral silence of science has, after all, revealed some of its less advertised aspects when the issue of production and disposal of corpses in Auschwitz has been articulated as a 'medical problem'" (Bauman 1991a, 29).

If scientific models and theories are indistinguishable from the technical apparatus that informed them or that will be formulated in their name and if ethical concerns are not accounted for by the technoscientific community, then what about public accountability and the responsibility of individual members of the technoscientific community?

Merton and his disciples developed the field of the sociology of science (whether of the Strong Programme of the Edinburgh School or one of its U.S. counterparts; see Pickering 1992) in order to account for the success and failure of the scientific ethos within its institutionalized formation in contemporary capitalist societies. As they and others, like Polanyi and Kuhn, all agree, the issue at stake is the power relations among constituents who have claims not limited to the general uses of knowledge but applicable to the particular fruits of technoscience.

The development and use of Zyklon B and the tests of its effectiveness in gas chambers differ little from the examination of the efficiency of ovens and chimneys for burning human bodies. Can the technoscientific community be spared the critique accorded political institutions because it is supposedly value neutral and therefore defenseless against the abuses of its products? According to Martin

Gilbert, SS Lt. Col. Dr. Wilhelm Pfannenstiel, professor of hygiene at Marburg University, and Kurt Gerstein, the chief of the Waffen SS Technical Disinfection Services, were asked to improve the killing technique that used Zyklon B. Their position and rank within the SS hierarchy committed them to participation in the killing of Jews because it was state policy. They may have agreed with the Nazi ideal of racial hygiene and thereby could have justified to themselves their involvement. But would their attitude and responsibility be different if they only worked in a remote laboratory rather than visiting the concentration camps, as they did (Gilbert 1985, 426–427)? Would a position in an academic or professional institution shield them from personally asking ethical questions?

These questions are amplified further, according to Bauman, when one considers the proactive orientation adopted by the technoscientific culture of modernity (at least in the sense of the potential perfectibility of the human condition):

> From the Enlightenment on, the modern world was distinguished by its activist, engineering attitude toward nature and toward itself. Science was not to be conducted for its own sake; it was seen as, first and foremost, an instrument of awesome power allowing its holder to improve on reality, to re-shape it according to human plans and designs, and to assist it in its drive to self-perfection. Gardening and medicine supplied the archetypes of constructive stance, while normality, health, or sanitation offered the archmetaphors for human tasks and strategies in the management of human affairs . . . Gardening and medicine are functionally distinct forms of the same activity of *separating and setting apart useful elements destined to live and strive, from harmful and morbid ones, which ought to be exterminated.* (Bauman 1991a, 70)

The justification for extermination seems odd when it comes from the long tradition of the European Enlightenment, which was supposed to overcome prejudices, hatred, and despotic power. But the rationale for extermination remains steadfast in the modern mind, whether in medical or gardening terms.

Leo Marx adds to Bauman's notion of the cultivation of the garden:

> A whole series of ideas we identify with the Enlightenment helped create a climate conducive to Jeffersonian pastoral [including] [n]ature as a universal norm[;] . . . the condition of man in a "state of nature"; and the simultaneous upsurge of radical primitivism (as expressed, for example, in the cult of the Noble Savage) on the one hand, and the doctrines of perfectibility and progress on the other. (Marx 1964, 88)

In his analysis of F. Scott Fitzgerald's *The Great Gatsby*, Marx writes:

> This hideous, man-made wilderness is a product of the technological power that also makes possible Gatsby's wealth, his parties, his car . . . The car and the garden of ashes belong to a world, like Ahab's where natural objects are of no value in themselves . . . the machine represents the forces working against the dream of pastoral fulfillment. (Marx 1964, 358)

So by the time modernity is challenged, it is from a variety of perspectives, such as literature and politics, and not exclusively by philosophical or sociological discourses.

The Modern Route to Perfectibility

The Enlightenment project of the eighteenth century in its striving for economic liberation, personal freedom, political equality, and educational progress was just as much based on the belief in human perfectibility as on the separation between religious or occult prejudices and scientific or secular beliefs. Though the shift from accepting the views and prescriptions of theologians and magicians to accepting those of the custodians of science had an enormous impact on European culture of that period, it also required that the confidence of individuals in their own capacity to think critically and improve their intellectual ability be bolstered. It was crucial to endorse the view that the human mind has the potential to expand its horizons and grasp the depths of knowledge on a multiplicity of levels. Whether in the works of Kant or Condorcet (1979), the trust in the development of the human intellect (and all the great advances it promised) was such that it fostered a belief in the potential for the progress of civilization as a whole.

Under what conditions would it be possible to enhance the development of the human mind? Though the human mind remains an amorphous entity, some material conditions, such as food, health, and literacy, are essential for its well-being. These conditions are at once of a private and public nature: The concerted effort of society as a whole must be guaranteed for each individual, and obstacles to these social efforts must be eliminated. Obstacles come in a variety of guises, not least of which are the "lives not worth living" (Proctor 1988, Ch. 7). The determination of who belongs in this category may vary from country to country and is therefore also historically contingent, but the legitimation of this process dates back to the Bible and the wars it sanctioned, as well as to later scientific texts of Charles Darwin, to the social Darwinists, and to the development and unfolding of the Eugenics movement in Western Europe and the United States. (This way of thinking, whether sanctioned by God or king, leads to so-called holy wars and wholesale expulsion, as the Jews have experienced since the destruction of the second temple in Jerusalem in A.D. 70.) As Robert Proctor so eloquently explains:

> The ideology of racial hygiene, at least before the mid-twenties, was in this sense less racialist than nationalist or meritocratic, less concerned with the comparison of one race against another than with discovering principles of improving the human race in general—or at least the Western "cultured races" (*Kulturrassen*). (Proctor 1988, 20)

In order to make the nationalist ideology of racial hygiene palatable, in order to reverse the decline in the birthrate and the increase in the hospitalization of the mentally ill, an appeal was made to science. As Proctor argues: "The appeal to scientific objectivity and neutrality served a dual social function." On the one hand, science legitimized "common perceptions and prejudices," and on the other hand, it was conveniently used to establish a set of criteria by which to rank races (Proctor 1988, 62). Legitimized originally by the Enlightenment, science has in turn become the legitimizing tool for views that extend (and at times inadvertently run counter to) the ideals of the Enlightenment. Once scientific legitimacy is granted to views concerning human traits and hereditary (i.e., genetic) characteristics (in Francis Galton's sense of eugenics, 1908), then it does not require an intellectual leap to move from classifying the physically handicapped and the mentally ill with the Jews, so that anti-Semitism is "medicalized" and bureaucratically justified (Proctor 1988, 194–195).

If racial hygiene is based on the view that the health of a community can be rationally controlled and improved, and if this view is widespread in its eugenic formulation among the most enlightened of nation-states (e.g., England, Norway, Sweden, France, the U.S.S.R., the United States), who legalized some forms of euthanasia (Proctor 1988, 285), then there should be no wonder that Nazi doctors and party officials embarked on the Final Solution to the Jewish problem. Their policy manifested the combination of "a philosophy of biological determinism with a belief that science might provide a technical fix for social problems" (Proctor 1988, 286).

The ideals of science were subverted, so it seems, in the use of medicine in Nazi hands; the political agenda of the Nazi party tainted the objectivity and neutrality of science and scientists (for more on German science, see Cassidy 1992, Ch. 21). Yet as Proctor aptly illustrates, the case of Nazi Germany is not merely one of the abuse of science as it is applied but, rather, a process that contextualized science as if it were operating in a political vacuum while deliberately (if inadvertently) fulfilling a specific political agenda. "The Nazis 'depoliticized' problems of vital human interest by reducing theses to scientific or medical problems, conceived in the narrow, reductionist sense of these terms" (Proctor 1988, 293). The Nazis had the help of willing and eager physicians and intellectuals who set the tone and the conditions for the rapid deployment of the instruments of science (Proctor 1988, 289). The eagerness of Nazi scientists had to do not only with patriotism turned nationalism but also with professional competition and with the help they expected to lend to Smith's "invisible hand" in opening job opportunities once Jewish scientists were eliminated (see also Cassidy 1992, 306 and Ch. 17).

Proctor continues to explain that "Nazi racial scientists advocated a 'medical-biological revolution' that would reorient science around racial values" (Proctor 1988, 291). They did so in light of previous theories, models, and practices from the whole of the Western world. Nazi scientists did not invent euthanasia or sterilization; they only extended their use to cover larger segments of the population with great efficiency. They cleverly confused the voluntary nature of death with dignity with the social expense of caring for the "unfit" so that they could garner a fairly widespread public support for their policies (Proctor 1988, 178). This is not to say that there were no prejudices against Gypsies, Jews, Communists, and homosexuals, as Daniel Jonah Goldhagen describes in *Hitler's Willing Executioners* (1996). It is to say, rather, that these prejudices could be cast in a light that transformed their irrationality (based on fear, envy, or whatever other emotional outbursts) into a rational process (based on science).

The consequences of viewing the world in terms of human perfectibility may seem harmless at first sight, for there is nothing more reassuring than to be told that by tending to one's intellect great benefits are bound to accrue (to the individual and to the community as well). The more attention one pays to health, environment, and forms of interaction, the more successful one will be, the longer one will live, and the more peaceful the world will become. A parallel scenario is portrayed in the case of the health of the environment (as we saw earlier). The metaphor of life as a garden has its unintended consequences just as does the metaphor of health and racial hygiene, according to Bauman:

> All visions of society-as-garden define parts of the social habitat as human weeds. Like all other weeds, they must be segregated, contained, prevented from spreading, removed and kept outside the society boundaries; if all these means prove insufficient, they must be killed. (Bauman 1991a, 92)

Who is to determine whether or not the Jews or the Gypsies or the homosexuals are indeed "weeds" to be exterminated once and for all? Is there something about modern culture that is condemnable because Nazi doctors, in the name of technoscience, experimented most abhorrently with human subjects? One could argue that medicine is not part of technoscience but, instead, the art of healing and therefore more prone to human folly and abuse than anything technoscientific (e.g., Delkeskamp-Hayes and Cutter 1993). Yet one must admit, with Proctor, that the depoliticization of medicine in particular and technoscience in general is a double-edged process: Under some circumstances the neutrality of science enables human emancipation in the name of truth, whereas under others it may turn out to enable human cruelty and destruction.

It may be useful to quote part of a letter written by Albert Einstein to his colleague, Max von Laue, written on May 26, 1933. In this letter, Einstein refuses to uphold a neutral position toward politics because such a position may turn him into an irresponsible citizen and perhaps also an irresponsible scientist. In this regard, then, he sets the stage for debates that raged within both the German and the U.S. technoscientific communities during and after World War II:

> I do not share your view that the scientist should observe silence in political matters, i.e., human affairs in the broader sense . . . Does not such restraint signify a lack of responsibility? Where would we be had men like Giordano Bruno, Spinoza, Voltaire, and Humboldt thought and behaved in such a fashion? I do not regret one word of what I have said and am of the belief that my actions have served mankind. (Quoted in Cassidy 1992, 302)

The Question of Responsibility

As is apparent from the previous sections of this chapter, no matter whether one's vantage point relates to the Enlightenment or to any other cultural construct and its contemporary manifestations, the very reality of Auschwitz and Hiroshima requires us to focus on the question of responsibility. This is not to say that there is a simple moral equivalent between the responsibility of the technoscientific community involved in the extermination of humans in gas chambers and the responsibility of the one that developed nuclear bombs. Each member of these communities bears witness and responsibility in different ways. Yet all of them follow the suggestions of the proponents of modernity who delivered their ideological promises through the rational institutionalization of scientific principles and political tenets.

Once a world war was underway, it became unclear to what extent principles and tenets could be upheld and by whom. Bauman puts the matter in the most extreme form, suggesting the possibility of distinguishing between technical and moral responsibility:

> Technical responsibility differs from moral responsibility in that it forgets that the action is a means to something other than itself . . . To put it bluntly, *the result is the irrelevance of moral standards for the technical success of the bureaucratic operation.* (Bauman 1991a, 101)

A bureaucratic operation may be that of a laboratory experiment, the provision of health care within or across national boundaries, or a war. Michael Walzer, for one, carefully examines the particularities of wars and the question of responsibility as it pertains to them. As mentioned earlier, members of the technoscientific community are in the double bind of not being able to apprehend the entirety of any of their projects in the age of modernity and of not being able to anticipate the uses (military or other) to which their activities will be put. This double bind resembles what Walzer terms the principle of "double effect" (first worked out by Catholic casuists in the Middle Ages): "Double effect is a way to reconcile the absolute prohibition against attacking non-combatants with the legitimate conduct of military activity" (Walzer 1977, 153).

Walzer critically evaluates this principle by outlining its rationale, beginning with the legitimacy of a particular act of war coupled with a direct effect that is "morally acceptable" (e.g., because an aggressor enemy is threatening the food supply of a nation). Then an "evil effect" is introduced as an exogenous effect; it is neither an end nor a means of accomplishing the legitimate effect; and this secondary effect is compensated for by the "good" one in a manner that is "justifiable under Sidgwick's proportionality rule" (Walzer 1977, 153). Walzer, for his part, is concerned with the utilitarian calculus that may prevail in such acts of war, and he therefore revises the rationale: "Double effect is defensible . . . only when the two outcomes [the good and the evil] are the product of a *double intention*: first, that the 'good' be achieved; second, that the foreseeable evil be reduced as far as possible" (Walzer 1977, 155).

If one were to apply Walzer's revised principle of double effect, one may find that direct responsibility lies with political leaders and not with scientists at all. For it is political leaders who identify the "good effects" in light of political considerations and justify them within a context different from the one preoccupying the technoscientific community. Moreover, the responsibility of political leaders extends beyond the accomplishment of good effects to the limitation of all anticipated "evil effects." However, how would these leaders be able to anticipate evil effects without the informed opinions of the leaders of the technoscientific community? Is it possible to ignore nuclear engineers and experimental nuclear physicists when deciding on the efficacy and consequences of an atomic bomb? Will the historical record suffice as a guide? Is the historical record itself not fraught with ambiguities and open-ended interpretations?

Michel Serres contextualizes these questions in general terms in relation to the "arrogance" of the proponents of scientific knowledge:

> May the aforesaid scientific knowledge strip off its arrogance, its magisterial, ecclesial drapery; may it leave off its material aggressivity, the hateful claim of always being right; let it tell the truth; let it come down, pacified, toward common knowledge. Can it still do this, now that it has vanquished temporal power and reigns in its place, a clerisy? Is there any chance of it still wanting to celebrate a betrothal between its imperial reason and popular wisdom? (Serres 1995, 5)

For Serres, the answer lies with a reconceptualization of philosophy as well as of science and science's appeal to popular culture. Though the scientific discourse had an emancipatory potential no different from that of the philosophical discourse, it seems to him that the scientific discourse has lost this potential, having become a means to political ends and public expectation. As he says:

> Philosophy is only performing its regular duties when it drills us in liberation. There was a time when science liberated us from certain slaveries and from darkness, there was a time when this discourse of circumstance was quite simply true. Sadly, the time has come when the sciences

are letting themselves get trapped in the customary subservience of groups who are looking only to perpetuate themselves as a group. (Serres 1995, 105)

To some extent, this formulation directly addresses the central theme of my book, namely, that now, by the end of the twentieth century, the ideals of science and its ethos have been compromised beyond recognition. Instead of searching for blameworthy candidates and exposing their follies here, I suggest uncovering the historical conditions that brought this situation about and that still guide our culture. If these conditions no longer apply, then we may need to overcome their trends in the next millennium.

It would be all too easy to lay all the blame of the unintended consequences of technoscientific developments on the shoulders of either political leaders or leaders of the technoscientific establishment. In either case, we would ignore the public's desire to shirk its responsibility for making choices, for negotiating a moral compromise in the face of fear and threat when the very survival of the human community is at stake. As the next chapter tries to illustrate, the public at large (and thereby its individual members) is responsible for setting the stage on which political and technoscientific leaders have to perform.

My focus remains on technoscientists and not on politicians because the dependence on credibility is fairly one-sided between these two groups of actors. We are more likely to find politicians claiming that a particular enterprise can be achieved because "scientists" say so than the other way around. Whereas we expect to be skeptical when hearing the pronouncements of politicians, we take to heart almost everything technoscientists say. One may claim that this is no longer the case and that technoscientists are now lumped together with politicians as the new icon to be targeted for public ridicule. Yet I challenge my fellow citizens to say this to the technoscientists who are attending to their leaking plumbing, their broken-down cars, or their head injuries. Our fundamental dependence on technoscientists must be acknowledged before it is criticized.

Moreover, I challenge intellectuals and critics of the technoscientific arena to play the role they have ascribed to themselves over the years. They need to recall the prerequisites that make their role possible and worthwhile: To know the subject matter before criticizing it; to recognize the point of view of the subjects so as to account for the complexity of their work; and finally, to acknowledge one's own fallibility as much as the fallibility of the targets. Under these circumstances, a different level of empathy may be warranted, and a different attitude may permeate the exchange between technoscientists and their critics. The intellectual can be a more effective critic when appreciating that critique is an interpretive work based on translation between discourses rather than a series of judgments based on absolute truths.

In this context, then, it seems advisable to follow Bauman's observation that intellectuals have shifted their roles from legislators to interpreters and as such stand in a different relation to political leaders and the public. As he explains, the legislative role of intellectuals under the modern model consisted of

> making authoritative statements which arbitrate in controversies of opinions and which select those opinions which, having been selected, become correct and binding. The authority to arbitrate is in this case legitimized by superior (objective) knowledge to which intellectuals have a better access than the non-intellectual part of society. (Bauman 1987, 4)

By contrast, intellectuals' postmodern strategy is interpretive:

> It consists of translating statements, made within one communally based tradition, so that they can be understood within the system of knowledge based on another tradition. Instead of being oriented towards selecting the best social order, this strategy is aimed at facilitating communication between autonomous (sovereign) participants. It is concerned with preventing the distortion of meaning in the process of communication. (Bauman 1987, 5)

Bauman goes on to explain that "between communities (traditions, forms of life) intellectuals are called upon to perform the function of interpreters; inside their own community, they are still to play the role of legislators of sorts" (Bauman 1987, 145).

If one adopts Bauman's qualified classification, then an outright indictment of technoscientists by intellectuals and critics is misguided. Appreciating public needs and cultural pressure for answers about one's survival as interpreters would enhance a principle of conceptual charity toward those whose practices are indeed interpretive as well. We may place a different set of expectations on members of the technoscientific community than on other academics, intellectuals, and cultural critics; but we should not confuse our different expectations with the interpretive roles all members of all communities are bound to be performing at the end of the twentieth century.

2

Public Expectations of Technoscience: From Truth to Immortality

The first part of this chapter explores how we have voluntarily brought ourselves into the modernist situation described in Chapter 1. It focuses on the need for control of both the social and natural environments through an authority one can believe in, an authority that deflects superstition, dogma, and manipulation. The authority of rational discourse and practice of technoscience, with its logic and ordered reasoning, with its explanatory and predictive powers, remains an appealing candidate to accommodate the public's need for stability, security, and trust, especially in the postmodern age.

I trace the quest for control through Ernest Gellner and Sigmund Freud, referring back to the ideas and principles of conduct of Epictetus, the Stoic of the first and second centuries A.D. If this control is placed in the hands of the technoscientific community and if this community adheres to Merton's ethos of science (as a community anyone can join), then the public may feel comfortable appealing to this community for its reassurance. If traditional obstacles to participation in control, such as lack of aristocratic standing or capital, are cast aside in the name of intellectual honesty, integrity, and fraternity, then the public's delegation of authority and power over the decision-making processes that affect its ability to control its destiny seems warranted.

I end this chapter with a reminder of a trivial yet significant factor regarding the enhancement of the technoscientific community's power and control over the destiny of humanity: Technoscience maintains its modernist appeal because if offers practical fruits everyone can enjoy. These fruits are neither metaphorical nor abstract; they are contextualized within modernist and capitalist ideals and models and are made available at a price to anyone able and willing to pay. The eighteenth century is still with us.

On Control

Control is propelled by fear of the unknown, but at the same time it leads to new forms of fear. Instead of being a point of termination, achieving control is an intersection point through which the human psyche and behavior traverse over time. Perhaps the concluding lines of Sigmund Freud's *Civilization and Its Discontents* may serve to connect the previous chapter with this one and the subsequent one:

> Men have gained control over the forces of nature to such an extent that with their help they would have no difficulty in exterminating one another to the last man. They know this, and hence comes a large part of their current unrest, their unhappiness and their mood of anxiety. (Freud 1989, 112)

To some extent, we are always looking for an explanatory principle with which to handle a great deal of data, multiple phenomena, and complex situations. While religious institutions gave us many such principles in the form of sacred texts and dogmas, human curiosity, if not impatience, impelled us to develop concurrent explanatory models that eventually came to be known as science. (On the motives for scientific explorations, see Stephan and Levin 1992.) When religious principles are cast aside, one worries that no explanatory model whatsoever remains, and therefore the growing appeal of science can be understood both in terms of its position as a replacement for religion and in terms of its partial success, which is commonly understood in the manifestation of technological innovations. However the choice between competing explanatory models is historically made, what remains relatively unshaken is the quest for some level of control of one's environment (in Baconian terms) and of one's life plan. Modern science embedded the quest for control in naturalistic terms and eventually added to them layers of theoretical principles and models. These were in turn used to justify one's choice of control mechanisms.

However much the Enlightenment ideals unequivocally supported the principles of logic, reason, and rationality, these principles were not well organized in general, nor were they powerful enough in particular cases, theories, and models. Bauman turns the tables on Freud and tries to tease out the scientific quest for order and certainty undertaken by Freud:

> It was in science, therefore, that Freud sought the court of appeal against the humiliation administered by a society which was neither good nor enlightened . . . Freud came to science as a rebel—even if, politically, he was moderate, liberal, mildly conservative and devoid of any sympathy for red flags and barricades. He needed to use the authority of science to lay bare the bluff of another authority whose verdict he wished to invalidate. He needed a science whose authority could be so used. He had to build such a science, virtually from scratch. As in the case of Kafka, "everything had to be acquired" as "nothing was granted." (Bauman 1991b, 170)

Some understand the search for explanatory principles, and the control over situations this search would enhance, teleologically and find a historical path, an epistemological trajectory, or a metaphysical insight with which to account for and accomplish this search. The residues of a teleological search are associated most commonly with the inspiration of Hegel, who sees world history and its ongoing progress, whether scientific or logical, proceeding exponentially toward the ultimate stage of a fully realized world Spirit. Others (from Kant to Marx and beyond) understand this search in reductionist terms that efficiently reduce a vast amount of confusing and at times even seemingly contradictory data into a simple layer, matrix, or principle. The reductionist move attempts to account for and accomplish the same goal as the teleological: to simplify so that one can better control one's actions and environment.

The quest for control is explained anthropologically by Ernest Gellner in terms of the three options open for the epistemological frontiers of humanity: order, chaos, and magic. The mechanistic view of nature in Gellner's sense assumes an ordered nature; an ordered nature is a "postulate" and therefore poses no logical or conceptual problems. By contrast, "the ghost," as Gellner calls the alternative view of nature, may opt for chaos or magic. "Chaos," he states, "is not an option. No society, no culture or language, either does or can exist, which operates on the assumption of a chaotic nature, of a world not amenable to conceptual order" (Gellner 1974, 124). "But," he continues, our society, which is based at least to some significant degree on the assumption of a regular and orderly nature, in effect a bureaucratized nature, amenable to control and manipulation and prediction, does have other frontiers which *are* inhabited: the frontiers with what may be called magical or participatory societies . . . Chaos is not an option, but *magic* is. (Gellner 1974, 125–126)

Given the horizon or continuum of choices every society makes in regards to its worldview (and its construction of an ontology, metaphysics, and epistemology), Gellner suggests that the borders of the ordered universe abut magic rather than chaos. While chaos is inherently "uninhabitable," the worlds of magic are inhabitable and as such provide refuge from the dissonance that may be felt under the strictures of an ordered worldview. When orderly explanatory models fail, magic is an alternative model that complements or supplements with its own explanatory power.

Serres disagrees with Gellner's conclusion that chaos is an uninhabitable concept for society, for he offers a mitigating process wherein philosophy would usher chaos into the scientific world through its conceptual creations and inventions so that the strictures of rationality would be less constraining:

We must introduce into philosophy the concept of chaos, a mythical concept until this morning, and despised by rationality to the point of being used nowadays only for discourses on madness. When one imports a concept that comes from science into philosophy, one has little bother, all the work is already done. An inspector of finished works, turned toward the past, the importing philosopher, retrograde, is a rationalist at minimal expense. He pockets the fame, the honor, the dividends without ever having paid the price. Cheater. The great hope is to import concepts scienceward, for practices yet to come. (Serres 1995, 98)

Serres's view, if understood as an invitation to account for and tackle chaos in ways that would make it more rather than less comprehensible, would complement Gellner's view and also support my claim that the quest for control is associated with and dependent on the need for explanatory and predictive models with which to organize societies (what John Dewey calls the "quest for certainty," 1929). This need or desire may be understood from a variety of perspectives: anthropologically, psychologically, epistemologically, politically, socially, economically, or religiously. No matter what prism one chooses, this quest is a thematic underlayment that can be traced historically as far as one wishes, and it finds current manifestations in the scientific revolutions of the modern era and in the Enlightenment ideals fostered by modern nation-states as substitutes for religious dogmas about divine designs of order.

Bauman's take on the Enlightenment is in contradistinction to the propaganda of truth, reason, science, rationality, and education:

> Enlightenment was, instead, an exercise in two distinct, though intimately related parts. First, in extending the powers and ambitions of the state, in transferring to the state the pastoral function exercised previously (in a way incipient and modest by comparison) by the Church, in reorganizing the state around the function of planning, designing and managing the reproduction of social order. Secondly, in the creation of an entirely new, and consciously designed, social mechanism of disciplining action, aimed at regulating and regularizing the socially relevant life of the subjects of the teaching and managing state. (Bauman 1987, 80)

Once we approach the psychological literature, whether that of the history of ideas (e.g., Baruch Spinoza's views) or that of professional psychotherapists (e.g., the work of Sigmund Freud and his disciples), we find that the narratives of control are enormously rich. The control of one's own appetites and desires clearly demarcates the formation of an ego that appreciates boundaries and is self-structuring. On the one hand, there is a positive element associated with the notion of control, broadly construed. On the other hand, there is also the contradictory sense that control of one's innate desires (e.g., sexual drives) may lead to frustration, anxiety, and pathologies harmful to the individual and to the environment. So the psychological view of the quest for and need of control opens possibilities for multiple interpretations.

Freud defines "civilization" as "the whole sum of the achievements and the regulations which distinguish our lives from those of our animal ancestors and which serve two purposes—namely to protect men against nature and to adjust their mutual relations" (Freud 1989, 42). In this context, Freud emphasizes the quest for and achievements related to human control over the environment, whether through fire, the production of tools, or the use of machines (Freud 1989, 42–46). The quest for control exhibited throughout the development of civilization as Freud describes it is expressed in terms of regulation when it comes to human relations. In this context, the community exerts power over individuals through rules of conduct, laws, and the entire justice system, which requires all to obey it (Freud 1989, 46–52).

Multiple interpretations of the notion of control and its practical manifestations make impossible an easy judgment about the merits or demerits of the concern with and focus on control. It remains unclear whether one should critically study forms of control in order to evaluate them, as illustrated by Freud when discussing civilization's "progress" from an adherence to the "pleasure principle" to an adherence to the "reality principle" and the sacrifices associated with this trade-off. Though Freud bemoans the frustration of the pleasure principle, he acknowledges the need for a growing population organized in a complex web of relations to accept the reality principle. As he says: "Civilized man has exchanged a portion of his possibilities of happiness for a portion of security" (Freud 1989, 73). This general assertion is explained in more detailed terms when Freud describes the internal structure of the psyche: "Civilization . . . obtains mastery over the individual's dangerous desire for aggression by weakening and disarming it and by setting up an agency within him to watch over it, like a garrison in a conquered city" (Freud 1989, 84). Freud's insight in this context is the parallel he draws between the development of civilization as a whole over the centuries and the development of an individual psyche over a lifetime: "We cannot fail to be struck by the similarity between the process of civilization and the libidinal development of the individual" (Freud 1989, 51). Whether the similarity is measured in terms of control or power depends on an appreciation of different historical developments and particular social formations, as some have done in regards to power (e.g., Dahl 1986).

In some respects, then, focusing on the quest for control in Gellner's or Freud's sense is a reductionist move: Can the sense of control, just like the will to power, provide an overarching explanatory strategy with which to comprehend everything? Reductionism as an epistemological strategy is dangerous and misleading, as we can see by considering the progress of eugenics in the past one hundred years, from the work of Francis Galton to the policies of Nazi Germany and the theories of the social Darwinists in the United States. The appeal of finding all the answers to our social and personal problems in our genes can turn into an appalling indictment against some people on behalf of others, so that some are deemed less worthy and are killed because of their genes. But perhaps there is a way to avoid the specter of reductionism and speak more broadly and openly about the quest for and accomplishment of control in Stoic terms. I limit my discussion here to Epictetus.

The Stoics claimed that there are things that are under our control (or up to us) and those that are not (Epictetus 1983, 11). Their bifurcation of the world can be seen as a reductionist move. But is it? When we speak of reductionism, we usually mean by it reducing multiple items into one (ignoring many subtleties in the process) and thereby establishing a hierarchy or a foundation of sorts. The Stoic move, though demarcating two sets of items, allows a multiplicity of items in the sets. So let me suggest that the Stoic demarcation, however problematic, is nonreductionist. Let me also suggest that it makes sense to speak of those things over which humans do and do not have control. What follows from this demarcation?

First, it does not follow that we all can agree on which items belong on which side of the ledger, for what may be deemed to be within human control by some may be deemed beyond human control by others. (Incidentally, the reference to a ledger suggests a particularly interesting and imperative role for philosophers.) Second, though the demarcation between those items under control and those that are not is blurred, it does offer a practical way of approaching one's life. That is, if a demarcation is accepted, however provisionally, a definite prescription can emanate from it. Third, an acceptance of a demarcation strategy recognizes in principle the limits and boundaries of human will and action, that is, the limits of what humans can conceivably control. As Epictetus says in his *Handbook*:

> So remember, if you think that things naturally enslaved are free or that things not your own are your own, you will be thwarted, miserable, and upset, and will blame both gods and men. But if you think that only what is yours is yours, and that what is not your own is, just as it is, not your own, then no one will ever coerce you, no one will hinder you, you will blame no one, you will not accuse anyone, you will not do a single thing unwillingly, you will have no enemies, and no one will harm you, because you will not be harmed at all. (Epictetus 1983, 11)

Part of the construction of the Stoic ledger is simple because various natural phenomena, like snowstorms and hurricanes, are deemed beyond human control. Other natural phenomena, like rain, have partially been set within human control with the advance of technoscientific instruments and techniques of aerial cloud seeding. So even natural phenomena, those attributable to God's creation or to the ongoing evolutionary process, are not fully contained within one side of the ledger. When it comes to the other side of the ledger, some recent theoretical and practical developments illustrate similar complications. Personal actions were traditionally deemed within human control, but genetic and environmental factors have now been shown to exert an influence no human can overcome. Where does it leave us?

Perhaps it leaves us with the following Stoic insight: "Do not seek to have events happen as you want them to, but instead want them to happen as they do happen, and your life will go well" (Epictetus 1983, 13).

The focus on the blurred demarcation is misplaced, for what is at issue is not what in fact we can or cannot control but what we perceive we should be able to control. In this respect, then, what transforms the formation of and interaction within society is the altered expectations of individuals: the reduced expectation that one has control over events and people in one's environment. Another way of saying this is to claim that the perceptions of control are coexistent and just as influential in control strategies that we succeed in as in those we fail to execute.

As Epictetus so aptly explains, what is under human control is the judgments humans make about events and people, about actions and modes of behavior, rather than these phenomena themselves. So attention ought to be paid to one's faculty of judgment, and its cultivation will illustrate to humans the difference between controlling one's judgment of a situation and controlling the situation itself. This way of thinking can be linked to later developments in the thought of Spinoza and Friedrich Nietzsche, for example, who both believed, albeit in different formulations, that one may never directly reach the "thing in itself." According to them, all we have access to are interpretations, personal perceptions, and subjective judgments.

Freud argued that most people find it difficult to differentiate in many cases between the fantasy one is "justified" in pursuing and the fantasy that one is not because most people have internalized and integrated the fantasy into their self-perception and the perception of their surroundings. Similarly, there is paranoia that is empirically justified and paranoia that is not. Still, the paranoid individual may have no access to the empirical justification until it is too late—for example, after the murder that was feared has been committed. With no evidence available until it is too late, the paranoid remains in the precarious position of claiming that someone is trying to murder him or her and being unable to prove it. Is this a fantasy or not? Does it matter when one thinks one's life is at stake?

The paranoid fearing murder must behave *as if* the murder, and therefore also its prevention, is within someone's control, knowing full well that the evidence—the intention of the murderer—will never be made public. This is the kind of control I was alluding to when I said that we tend to behave *as if* we control situations, regardless of whether they are empirically understood to be on the side of the ledger that warrants control-like behavior. Put differently, it really does not matter how carefully the control ledger is set up since humans are constantly appropriating items from the "out of one's control" category and moving them into the "within one's control" one. So does this mean that the Stoics' prescription for happiness is useless? Yes.

If the human condition—a category different from human nature— is such that it induces us to shift as many items as we can find into the category of controllable situations, then Stoic prescriptions remain hollow. They say one should worry only about those things within one's control and let go of those that are not. But if it is unclear what is and what is not within control, then how does one know how to approach a situation? Must one size up a situation and say: "Oh, this is outside my control," and then walk away? Would it not be up to others to challenge that and say: "But wait a minute, you are evading the situation, you can do something about it!"? And if they said this, what should be the response? Should their challenge be dismissed? Would their challenge thereby dissipate? In short, is there an inherent and objective logic to the situation regardless of opposing subjective interpretations?

These questions about one's control and responsibility are related to whether one sees the universe as ordered and to one's view concerning the distinction between an ordered natural and social world. Though I elaborate below on the case of the Manhattan Project, it may be useful to recall that Werner Heisenberg, one of the preeminent nuclear physicists in Germany before and throughout World War II, demonstrated a keen concern for an ordered reality, an order he expected to permeate the political arena as well as the world of inanimate objects (Cassidy 1992, Ch. 24). Heisenberg craved a level of order and stability in his life perhaps in Freud's and Gellner's senses because of the twofold pressure he felt: the pressure of Nazi ideology's influence on German science (in lieu of the theories of Jewish scientists, such as Einstein) and the pressure of an ongoing critique of Niels Bohr's Copenhagen Interpretation, to which Heisenberg subscribed (see, e.g., Cassidy 1992, 391). The hope for the neutrality of science and of one's personal beliefs was no longer an option under the Nazi regime.

In this context, it should also be noted that an acceptance of the view of an ordered universe has some unintended consequences regarding one's sense of responsibility. Heisenberg has been accused of saving "physics [but] not . . . physicists" (Cassidy 1992, 483). The controversy about his guilt or innocence is still debated (see a brief summary in Cassidy 1992, 424–426). His excuse, as David Cassidy reports, has to do with his view of the inevitability of historical forces, against which the individual, however influential, can do nothing. He tried to help some close associates but failed to make the weight of his power felt beyond his immediate circle of friends (Cassidy 1992, 424–426).

Across the ocean and across many additional cultural and political divides, a similar predicament (and its attendant sense of responsibility) has been articulated in the situation of the development of the nuclear bomb by a select group of technoscientists at Los Alamos. (Cassidy draws this parallel as well; Cassidy 1992, Ch. 26. Incidentally, the term "technoscientists" fits them well because however scientifically trained, they had to perform engineering tasks, devise experiments, and participate in all phases of laboratory work.) In his fare well speech to the Association of Los Alamos Scientists on November 2, 1945, J. Robert Oppenheimer made the following comments:

> There are a few things which scientists perhaps should remember, that I don't think I need to remind us of; but I will, anyway. One is that they are very often called upon to give technical information in one way or another, and I think one cannot be too careful to be honest. And it is very difficult, not because one tells lies, but because so often questions are put in a form which makes it very hard to give an answer which is not misleading. I think we will be in a very weak position unless we maintain at its highest the scrupulousness which is traditional for us in sticking to the truth, and in distinguishing between what we know to be true from what we hope may be true. (Smith and Weiner 1980, 324)

Though suffering under the predicament of the situation, Oppenheimer beseeches his fellow workers to have faith in their trade and to continue to uphold the tenets of science as he knows them:

I think that we have no hope at all if we yield in our belief in the value of science, in the good that it can be to the world to know about reality, about nature, to attain a gradually greater and greater control over nature, to learn, to teach, to understand. I think that if we lose our faith in this we stop being scientists, we sell out our heritage, we lose what we have most of value for this time of crisis. (Smith and Weiner 1980, 325)

To understand Oppenheimer's cultural background, one may need to refer back to the Stoics. Epictetus hoped to accomplish a fantastic feat: to provide a handbook for better living. If one accepts the demarcation between those things that are under one's control and those that are not, and if one focuses attention only on the things under one's control, then one can expect to radically reduce frustrations and anger. Who can resist such an attractive prescription? The Stoics' appeal rests on the simple assumption that one can distinguish between those things that are and those that are not under human control. But what happens if, as proposed above, it is impossible in principle and in practice to maintain such a clear distinction? In that case, one would be hard pressed not to fall into the predicament of believing that more things are under one's control than in fact are. This would lead one to attempt controlling uncontrollable things and thereby to remain frustrated, angry, and unhappy. Similarly, one may believe that fewer things are under one's control than in fact are, and therefore one is bound to fail to attempt controlling controllable things and thereby to remain just as frustrated, angry, and unhappy as in the previous case.

Remember, though, that the goal is happiness even when defined minimally as the avoidance of angst. If the Stoics are right about the strategy for achieving happiness even though they are wrong about the constitution and interpretation of the control ledger, then it makes sense to insist on being able to distinguish between those things that are under human control and those that are not. Hence, it also makes sense to insist that there are criteria according to which the distinction can be accomplished. These criteria may be useful in ascertaining the logic of the situation so as to adjudicate among multiple interpretations.

Technoscience as an Invitation

As we move into the twenty-first century, we cover old ground, searching for past insights, so as to guide our future. Our journey reflects an acceptance of the human condition in terms of the quest for some level of control, that is, in terms of an ongoing negotiation with one's personal frailty, the demands of others, and the circumstances of the environment. Instruments of control extend from the operation of the mind to the sets of tools we have invented and perfected over time. In this context Francis Bacon comes to mind, not so much because he revolutionized a particular area of study but because he insisted that superstition and speculation must be replaced with detailed observations and experimentations. It is often said that when Bacon equated knowledge (i.e., scientific knowledge) with power (i.e., intellectual power), he meant by this the conquest of nature, its exploitation, and abuse. But Bacon also understood that nature, in order to be controlled, must be obeyed (Bacon 1985, No. 3), especially since he realized that "Man, [is] the servant and interpreter of Nature" (Bacon 1985, No. 1). What does obedience mean in this context?

One can speak about the manipulative move from obedience to mastery, so that every act of adherence is potentially dominating and oppressing. One can also speak about obedience and control dialectically (in Hegel's phenomenological sense of master and slave), so as to show the ambiguity and exchange of power roles and relations. Finally, one can speak about obedience to nature literally, so as to acknowledge the insurmountable power of nature over the frailty of humanity. In any of these three senses, it becomes clear that "nature" and "science" are categories that not only define each other but are also test cases for the quest of control.

Modern science is supposed to identify some rules and regulations, some order and rational structure, and some regularity (even if only probable) on nature. Nature is no longer God's creation or the mystical reflection of divine beings who speak to humans through natural phenomena (such as rains and storms, droughts and floods). Instead, nature can be comprehended rationally; it can be modeled in a way that appears to adhere to some order; it can make sense to humans. By having nature make sense, or in other words, "obey" certain laws, we have forced the issue of control, for is not the very organization of knowledge a matter of cognitive control? Of course, one would hesitate to jump from this kind of control—a simple cognitive exercise that remains bound to human-made rules and linguistic techniques of expression—to the kind of control that claims to control natural phenomena themselves: an absurd exercise in commanding a set of variables to coincide at will.

It is important to emphasize this point: Unlike some social constructivists who are concerned with the sociology of scientific knowledge (ssk) to the extent that all scientific activity is reduced to and explained by the social context of the community of scientists (e.g., the Edinburgh School Strong Programme), one can refrain from equating the control over models of nature with the control over nature itself. I know that what I just said is at once trivial and would be deemed a misrepresentation of the social constructivist agenda: The issue is the framing of nature through models so that what gets counted as nature to begin with is controlled cognitively. But this sense of control is much weaker in its claims since it admits some form of realism: There are phenomena whose ontological status either predates humans or is not codependent on human recognition and interpretation.

Scientific models, theories, principles, and hypotheses play right into both the strong and weak senses of control over nature. Any use of them collectively or individually is itself a recognition of the role they play: They mediate between humans and nature. Mediation, though, is a tricky concept, for it can be understood passively and actively: Either humans prefer to be shielded, distanced, even withdrawn from a direct interaction with their surroundings, or they prefer a direct engagement that changes the situation under consideration. In either case, it is reasonable to transform the strong constructivist argument into a weaker one, one that acknowledges the active role human cognition plays in determining the contours of nature without thereby denying that there is something out there whose contours are being defined. Another way of phrasing the issue is to say that the strong version is too strong, whereas the weak version needs to be strengthened a bit.

Here is another way of explaining my concern with overly strong constructivist versions: Only a small technoscientific community is able to uphold, sustain, and defend such a view. The weak view of the way in which technoscientists construct their models for the purpose of explaining and predicting natural phenomena is much more accessible to the general public. The more accessible an explanatory model, the more appealing it is to a large section of the public; the more appealing the model, the more comfortable the public is with it; and the more comfortable the public is with a specific model, the more faith it has in it. But my concern is not only for broad-based accessibility for the sake of building public belief systems. I am concerned also with the question of entry barriers set up for those interested in the techno-scientific community. If technoscience is constructed incomprehensibly, then structurally technoscientists replace shamans, witch doctors, and members of the clergy, whose privileged position is sanctioned by divine powers that demand obedience rather than critical engagement.

Moreover, if technoscience preserves a weak version of social and cognitive constructivism, illustrating time and again human folly and fallibility, then it becomes more inviting to aspiring generations of potential technoscientists. Since no one has all the answers about nature (because of lack of cognitive power and control) and since every technoscientific model and its practical application is bound to be challenged and revised (inter- or intragenerationally), then everyone can participate in the technoscientific endeavor. Perhaps "everyone" is an exaggeration, for some training is required. And here I would like to make a plea for the ideals that guided the nascent scientific community of three hundred years ago and for Karl Popper's sense of scientific progress as opposed to the structural descriptions and prescriptions formulated by Michael Polanyi and Thomas Kuhn.

Though the Stoics suggested we refrain from endeavoring to control those aspects of our lives that are beyond human control, and though we may have learned to agree with them that humans can never expect to control as much as they aspire to control, it still remains an open question of what to make of this situation. If Stoicism is not a recipe for inaction in the name of fatalism, then it is a recipe for responsible action in the face of a reality parts of which remain beyond our control. Our scientific ancestors understood the human predicament and expressed it in a variety of ways: the image of Frankenstein's monster, the image of Faust's fateful compact with the devil, and the image of technology being out of control. The balancing acts technoscientists must engage in, the thin line they must walk, on behalf of the rest of society puts the burden of responsibility on their shoulders. And as a group responsible for the well-being of the rest of society, this group feels at once privileged and overwhelmed.

In the name of keeping the cranks outside the respectable domains of the technoscientific community, its leadership raised high barriers and set strict gatekeepers. The gatekeepers do their work well when they allow into the community only those who deserve to be let in, while keeping all the others outside. The criteria for admittance are determined by the leadership in light of the community's prior accomplishments. That is, the criteria reflect the already established dogma of the community; the past informs the future. Though this practice is indisputably useful for "learning from experience or from errors" (Berkson and Wettersten 1984), it hampers potential deviations from the past. Put differently, if the past is the only guide for the future, then one may forever be cognitively stuck in the past.

Supporters of the gatekeepers and the criteria according to which they fulfill their role argue that the past is an adequate guide for the future because it prevents the technoscientific community from repeating past mistakes while allowing it to enjoy already established insights (see, e.g., Kuhn 1970). Opponents of this practice argue that the past can only be an incomplete guide because it cannot foresee what has not taken place yet (see, e.g., Popper 1959). Of course, they argue, we need to know about the past, but we should not be committed to it; on the contrary, we may be committed to overturning the past altogether. So gatekeepers may turn out to be inadvertently too zealous in arresting changes, revisions, or progress in general.

The plea to return to the original intent of the Royal Society of London and its later instantiation with Popper is simple: Let anyone through the gates who has something scientifically interesting to say that can be replicated (whether an observation or an experiment). Popper's proposal for conjectures and their falsification through rigorous testing is an invitation for a candidate to be as bold as he or she can imagine, without having to guarantee in advance that the hypothesis in question will be successfully verified or falsified. This sounds almost too good to be true. Can anyone propose any hypothesis whatsoever? Would it not be a prerequisite that one know in advance that the hypothesis makes sense, or at least enough sense to be tested in principle if not in practice? How could one know that much in advance? Must one not be already a member of the technoscientific community to venture such a courageous feat?

Having raised a set of questions about the conditions under which speculation can be considered scientific (warranting testing), it seems that the offensive role of gatekeepers can be ignored. If indeed no one would be silly enough to venture a conjecture out of thin air, so to speak, then the threshold for the activities of technoscientists renders gatekeepers superfluous: The threshold is formidable enough that the gates can remain open and without guards. But is the threshold also low enough to permit entry to those who may be too intimidated to enter freely into the technoscientific community? This question has been answered to the satisfaction of the scientific community and some of its critics, as mentioned in Chapter 1 in relation to Merton's formulation of the ethos of science. But what about other critics? Is it not the case that the technoscientific community only provides provisional answers and portrays a partial picture of an ordered universe? Has not the history of science itself been instrumental in overthrowing its own claims for permanent certainty?

Though the public expresses a willingness to consume whatever is produced by the technoscientific community and though that community enjoys the financial support and admiration of the public, this mutual reliance (almost codependence) should not obscure the difficulties with which both the public and the technoscientific community face their inevitable disappointment over the control of their environments. The use of myths, images, and metaphors (about knowledge, certitude, control, and happiness) may be a necessary stage in the development of a civilization, but it should not be taken as confusion in the face of an incomprehensible reality in perpetual flux.

The Myths of Technoscience and Capitalism

Instead of rehearsing the argument about the intimate relation between science and capitalism in general and using as the major example the Industrial Revolution in Europe and the United States (see, e.g., Toynbee 1956), I will discuss a particular aspect of both discourses and practices. Capitalism and technoscience are perceived to be egalitarian because they both employ the same myth, one whose status is upheld to this day. Let us begin with capitalism. Adam Smith tells us that though exchange is motivated by self-interest, it is undertaken voluntarily; it is free exchange. Smith easily argues for his "economic model" because he rides one of the moral waves of the Enlightenment: All humans are equally free and as such should be treated as ends and not as means (Smith 1937).

However many revisions this doctrine underwent over two hundred years, it eventually made its way into the late twentieth century under the title of Contestable Markets Theory (Baumol, Panzar, and Willig 1982). In short, this theory says that we have perfect (i.e., free and equal) competition whenever anyone can contest the players within a particular market by joining their ranks. The reigning myth, in case it is unclear, remains unchanged and is used as the yardstick for regulative intervention: We are all free to compete, and we are all equally free to do so. One recent modification has been to replace the notion of "equality" with that of "equal opportunity," but the guiding principle does not change (see, e.g., Sterba 1984). Unless this myth were upheld, one would be hard pressed to find so many people working so hard and being so motivated to improve their lot. They must believe that there is a way for them to compete, or they would not remain engaged in what would be an exercise in futility.

Myths are luxuries without which humans would not survive or at least attempt to survive. The driving force of capitalism was not the steam engine but, rather, a myth that united equality and freedom. This myth was so powerful that it overcame the power of tradition; it remains powerful in its ability to forestall the demise of capitalism. While Marx paid attention to the material conditions under which capitalism emerged and under which it would eventually buckle, we must follow more seriously not Max Weber's specific sense of the "spirit of capitalism" (Weber 1958) but the general and pervasive force of the myth of capitalism.

I would not link here the spirit of capitalism only to its Protestant corollary, a parsimonious, productive, and therefore ethical way of life. Instead, I would return to the human predicament already marked and announced by various thinkers from John Locke to Jean-Jacques Rousseau: We are born free and equal in the eyes of the Lord (or Nature, or both in Spinoza's sense), and we live a life of coercion and inequality. How is this possible? Who is the culprit? Is this fair? What should be done to remedy the situation?

Plato had no such problems; for him the issues of equality and freedom were contextualized so differently from the way they were by the figures of the Enlightenment that one would nowadays not call his views either egalitarian or liberal. For Plato (in the *Republic*), perhaps even more than for Socrates, there was a self-explanatory (i.e., natural) hierarchical ordering of the universe so that humans fit their predestined roles and function within a society in perfect harmony and without resort to radical reforms. For Plato, people are equally qualified for their roles in society, not equal in some abstract and absolute sense. They are likewise free to pursue their allotted tasks as best they can, not free to pursue any task they choose. To simplify in the extreme, Plato's views of equality and freedom are relativist.

Are the Enlightenment's figures mistaken in their insistence on the absolute nature of equality and freedom? Perhaps we should begin with this question so as to respond to the questions raised earlier about being born free and equal and living a life of coercion and inequality. For the myth of capitalism to succeed, that is, for it to be a myth people believe in, uphold, and fight for, it must be a myth about absolute values. If it were to portray equality and freedom as relative, then its rhetorical and persuasive power would be lost: It would no longer be a myth but a description of reality.

The myth of capitalism is the engine that drives capitalism. People believe that despite all odds and personal circumstances of birth, they have the chance of being just like anyone else, of being accorded the same opportunities as anyone else, and thus of being able to succeed like anyone else. Because of this deeply held belief, counterexamples, such as the systematic failure of some segments of the population to reach parity with others, are relegated to the personal, not the public, domain of explanation. If one is poor it is because of laziness, namely, because one actually chose to be poor; by working harder and applying more diligently to the pursuit of success, anyone can alter the situation and become rich. According to the apologists of the myth, the problem is not with capitalism as a system of power relations but with the individual who is part thereof.

Having a powerful backdrop in the form of a myth against which all players have to adjust their performances on the stage of capitalism ensures a coherent vindication of the system and a relentless condemnation of all those who fail to struggle for the ideal. But is this not exactly what also ensures the continuous motivation of people to work hard for themselves and for society as a whole? Is this not the most effective way of providing incentives (however elusive they may turn out to be) for people? Besides, are there no examples to prove that this system in fact delivers the promise that its myth sets up in the background? Have we not heard many rags-to-riches stories?

It is hard to argue against these stories of success, because in the back of our minds there is that little dream that one day our fate will be more like the fate of those who succeeded than of those who failed. In the western hemisphere (and perhaps by the end of the twentieth century more globally), we have this dream engraved in our minds since childhood by diverse forms of socialization (i.e., indoctrinations). But this is not the point I am trying to make here. To formulate yet another class-based Marxist critique at this stage of our cultural deconstruction is too little and too late. Instead, I want to emphasize how the capitalist myth parallels the technoscientific one, and how both are at once most appealing and most dangerous. They are appealing because they offer hope for those who may otherwise feel hopeless; they are dangerous because they promise something they cannot deliver and are therefore bound to frustrate, disappoint, and provoke anger.

The technoscientific community tells us that anyone can join its community, that we are all equally and freely invited. It is empirically true that people from all walks of life, so to speak, have successfully become members of this community. It is also true that the technoscientific community could afford refuge for those who were marginalized by the social and economic conditions pervasive in the capitalist system. So in these respects, the technoscientific community's myth of equality and freedom seems more honest in its presentation than the myth of capitalism. But before we conclude with this statement, let us recall some other factors that explain why this myth, though appealing, is also dangerous.

Two important factors endemic to the technoscientific community vitiate the credibility of its myth of equality and freedom. On the one hand, it remains thoroughly a community, an institution with rules and regulations, with a hierarchy and leadership. As such, it can never mitigate its inherent structural inequality and inevitable coercion. As Kuhn (1970) aptly reminds us, there are good reasons to require young people to follow the rules of their elders, to master the materials with which they will operate and experiment. Only after years of initiation will the new member be able to make a contribution, either to solve yet another puzzle within the paradigm or to recognize the antinomies of the paradigm and suggest a paradigm shift.

On the other hand, the technoscientific community does not exist in a setting of its own design: It remains entrenched and indebted to capitalist modes of production, distribution, and consumption. Hence, there is no sense in which the power relations of the larger system do not interfere with and inform the power relations of this particular community. I would not say that the scientific community is a microcosm of the capitalist community but, rather, that it cannot avoid some of the strictures of that larger context. These strictures, the way I understand them, are pervasive and unrelenting, as Paula Stephan and Sharon Levin (1992) carefully document. Technoscientists are not free to choose what project to work on, nor are they equally funded for their projects (see, e.g., the Society for Social Responsibility *Newsletter*s). Funding has become the driving engine for technoscience and inter alia has become the driving force behind the organization and power relations of the technoscientific community.

When the parallel myths of capitalism and technoscience are derided as untrue or dangerous in any sense, it seems as if there is a plea to banish them from our culture. I wish to make a different case here. First, I assume that myths are necessary components of every culture because they enhance a sense of unity and conformity among the individuals in the group. Second, I also assume that no matter how myths are originally set up, they eventually are revised and reinterpreted in the image of the age that deploys them. Third, I finally assume that though myths are important components in determining how cultures operate, they remain backdrops to which one may appeal but which one may also routinely ignore. Therefore, their functional importance remains dialectically ambivalent.

All of these assumptions lead me to conclude that public interest in technoscience, public trust in the technoscientific community, and public awe in the face of technoscientific feats all derive their potency from an adherence, however oblique, to the myths of capitalism and technoscience. These myths have the power to slow down a decline in faith in technoscience and a renewal of faith in religion and magic. This chapter tries to explain why it is that after so many generations of frustration with and concern over technoscience, there is still so much public support for this community. The explanation that tells us that the public is duped by technoscientists and that they, in turn, are good manipulators of public opinion is not a reasonable one. No matter how self-serving technoscientists are, they could never be able to systematically control the public that controls them; they could scare the public at times, intimidate the public at other times, but on the whole they would have to respond to public outcries and concerns, fears and hopes.

So instead of positing a situation in which a reluctant public supports a cunning technoscientific community, I propose the following scenario, theory, or hypothesis: The public likes the technoscientific community not only because of all the fabulous fruits and toys it has produced over the years but also because the public believes in this community's myths and its inherent values; the public considers this community as a viable instrument with which to control the Stoic control ledger. Technoscientists themselves may grow cynical after a few years in laboratories, the academy, or government agencies. The public, on the other hand, still remembers the myth that supposedly guides this community, the myth about equality and freedom, about being able to join at will and without unfair obstacles a community dedicated to the study of the secrets of nature. The public, then, expects redemption from the technoscientific community. Redemption, I may add, is understood here in the fullest sense of the term, for there are too many pervasive disappointments surrounding the public sphere, and the appeal to technoscience is a last resort after the disappointments of various "good books."

Moreover, the public may have become disillusioned with the myth of capitalism because of the instances of injustice, cruelty, exploitation, and alienation that undermine, refute, and falsify the myth of equality and freedom that enshrines capitalist modes of production, distribution, and consumption. If the capitalist myth falls short, is shown to be not utopian but outright false, then perhaps the counterpart myth of technoscience can make up for it. Perhaps capitalism, having given up on the credibility of its own myth, needs to appeal to the myth of technoscience in order to bolster its own myth (since each myth is a mirror image of the other).

There are cultures in which a leadership sets up a myth in order to fortify its modes of operation and legitimize its institutions; the technoscientific community, however, does little of this, contrary to what Merton (1973) may argue. Instead, the capitalist community, that is, the culture surrounding the technoscientific community, fortifies the myth of equality and freedom in order to restore a credibility these values may have lost over years of exploitation and alienation. I am not saying that the technoscientific community itself and its leadership are no longer responsible for perpetuating their myth and its variants; rather, I am arguing that the perpetuation of the myth is supported wholeheartedly by numerous constituencies outside that community.

So when we come to judge the technoscientific community and challenge its practices, we should first take notice of the external pressure put on this community for good reasons and by well-meaning people. There is something eternally appealing about freedom and equality: These two values at once conform to and defy a divine design. Even when institutionalized religions tell us that in the eyes of the Lord we are all equal and that predestination is no hindrance to free will, there is still a God out there, a Father who controls and dominates, who plays all the Freudian tricks on our subconscious. When Nietzsche kills God and frees us from our Freudian shackles, we believe that freedom and equality are now really ours to actualize. If capitalism falls short of this ideal situation, then perhaps technoscience can fulfill our dreams and hopes. What happens logically and psychosocially if technoscience falls short of this ideal?

The horrors of World War II, as we have seen described obliquely in Chapter 1, are sad reminders of why we cannot afford to lose hope of the technoscientific ideal and the Enlightenment myth that supports its operations. If we give up on technoscience, are we doomed to enter a vicious cycle that will take us back to religion and then to capitalism all over again? Have we fallen into Dante's Inferno, only to realize that there is no way out? Perhaps because Dante's Inferno haunts our imagination, we cling feverishly to the myth of technoscience and to the promises of the technoscientific community: We long for an island of sanity and reason, an island of reasonableness and prudence, an island of equality and freedom of thought and action. Admittedly, this is not the primitive island on which Robinson Crusoe and his man Friday arrived; it is an island in the midst of the wheels of commerce and industry. Yet, as so many outside of the technoscientific community would insist, this is an island nonetheless, and we wish to keep it that way: a sign of better things to come, a utopian vision!

With the myths of a utopian future and some accounts of horrible dystopias we approach the next chapter. In it, I will examine the inherent dread and excitement that inform the work of some of the members of the technoscientific community. I will account for their sense of their work, its motivation and fruits, primarily as expressed in plays and in group discussions as those reported from Farm Hall and the Manhattan Project. Perhaps they are neither definitive nor fully illustrative of the pangs of anxiety I wish to trace here, yet they will suffice as expressions that ought to be heeded by a society whose expectations are too high and unreasonable.

3

Ambiguity and Anxiety: The Making of Human Anguish

The first part of this chapter traces the human quest for order and the traditional gratification of this desire through religious doctrine (this parallels my examination of control in chapter 2). I then argue that the technoscientific community continued in this tradition to accomplish a similar goal of providing an ordered conception of the universe. Yet unlike the religious community, the technoscientific community appreciated and acknowledged, over time, the inherent and inevitable ambiguity of its trade. Ambiguity leads to a mode of thinking that is in constant search of additional alternatives and answers to given problems. It is this sense of critical self-evaluation that has marked both the acceptance and the success of technoscience all along. Ambiguity, then, forms the basis of the development of the technoscientific discourse.

The second part of this chapter moves from ambiguity to anxiety and argues that anxiety is the necessary (if not sufficient) fiber to guarantee "progress" and even the vitality and survival of the technoscientific community. Without the ongoing tension of the inconclusiveness of the technoscientific project, this project would come to a standstill, with no individuals or groups being interested in pursuing yet another unresolved anomaly.

The third part of this chapter suggests that there is a deep anguish as part of technoscience, an anguish that its members attempt to either cover up or ignore and that some, like Gerald Holton, characterize as "despair" (1996, 156). If, for example, more attention had been paid to the anguish of technoscientists during and after World War II, it would have been clear that anguish is not the exclusive domain of art and literature. This does not mean that the horrors of World War II would have been averted, but attention to technoscientists' anguish may have forced military and political leaders to reconsider their wartime proposals.

Technoscientific Ambiguity

When something is ambiguous, we believe it to be either obscure or having more than one meaning. This is true whether we speak of the most mundane facets of our daily lives or of nature as a whole. We suffer a certain discomfort when we feel surrounded by ambiguity; we like clarity and order, to know for sure what game we are playing and by what rules. Since our very existence in this world is overwhelming and frightening, any level of reassurance helps ease our daily concerns.

Ambiguity arises in our daily life and our perceptions of the world not only because we ourselves do not know what to make of this world but also, and even more so, because we receive more than one answer to the questions we ask about this world. The great influence and appeal of religious doctrine has always been its unwavering view of the world. I would venture to say that beyond the general concern for bringing together individuals into groups, religion—any religion, whether Christian, Jewish, or Muslim—understood its major task as easing the concerns or alleviating the fears of its members. Whatever a church's doctrine says, it must say unequivocally, providing absolute order in the face of chaos and easing the anxiety of its adherents.

This does not mean that there have been no heretics, that there have not been those who opposed the church's doctrine. But those who did oppose the church were cast aside as heathens, heretics, and sinners. All churches made sure that their congregations knew the fate of these outsiders: They were excommunicated and doomed to hell. However well these outsiders may do in this world, the ultimate punishment and revenge is reserved for the afterlife, when justice will eventually be done. The beauty of this scenario is that it remains untestable in principle, so that one must accept or reject this view on faith alone. (In this context, Popper's notion falsification as a test of scientific validity comes to mind.) Two related issues come forth from this brief description: First, the concern of a church is with an ordered universe, and second, whenever an alternative view emerges to challenge a given order, it must be discounted vigorously.

As for the first issue: Secular thinking of the late twentieth century at times forgets what religious doctrine is about, for it is not limited to saving souls from hell. On the contrary: it is just as concerned to provide an overall cognitive umbrella under which all ideas can shield themselves from unforgiving critical rains (reflect, for example, on Giordano Bruno's burning at the stake in 1600). Ideas, churches understood all too well, are very tender and sweet. They appeal to our tastes only insofar as they do not come under attack. When they do, they may be displaced, replaced, or even disappear. And then what shall one do?

So, in defense of religion, one must acknowledge the care with which churches undertook the construction and preservation of ideas, ideas about the creation of the universe, about the order of the world, and about interaction among humans. Ambiguity breeds doubt; doubt breeds confusion; and when the mind and soul are torn in many directions under conditions of confusion, people may not only lose their bearings and be frightened, but they may become divided. When divided, there is no telling what they will do: They may unite to fight the evil gods of nature, or they may fight each other to get the "right" answer. In either case, there is real danger that harmony on earth will not be achieved and that the very survival of the human race will be threatened.

The point here is not so much in defense of this or that particular church doctrine; I find them all suspect and inadequate. Instead, I wish to explain why it made and still makes sense not only for religious leaders to espouse their doctrines but also for the followers of these doctrines to appeal to them and to accept their authority. It is not because they suffer a lack of knowledge and pervasive ignorance that people become believers and disciples but, perhaps, because they have stared at the existential abyss that lurks behind every boulder and realized that the only way they can avoid their personal fall and the loss of their entire community is by adhering to something with absolute and universal explanatory power. Freud quotes Goethe in regard to the connection between religion and the worldviews espoused by science and art: "He who possesses science and art also has religion; but he who possesses neither of those two, let him have religion!" (Freud 1989, 23).

When the technoscientific community began to establish itself, with the success of the scientific revolutions of the seventeenth and later centuries, its first order of the day differed little from that of the established churches. It wished to organize, explain, and predict literally everything under the sun. Theories and models were comprehensive and pervasive: If true, they left little room for ambiguity. Modernism and the enlightenments of the eighteenth century were (and in many respects are still) movements whose appeal rests in their promise to reduce ambiguity. Their appeal to the modern mind and heart is the replacement of God with Nature (perhaps in Spinoza's sense), a concept seemingly less dependent on belief and faith and more on direct observation and intersubjective experience. Having said this, I do not mean to lump together religious doctrine and technoscientific theories of the past; instead, I mean to illustrate that their intent was similar: to assuage the human mind and soul in regard to daily activities, to ensure some stability and security, a sense of comfort and content. As Bauman says:

> We can think of modernity as of a time when order—of the world, of the human habitat, of the human self, and of the connection between all three—is *reflected upon*; a matter of thought, of concern, of a practice that is aware of itself, conscious of being a conscious practice and wary of the void it would leave were it to halt or merely relent. (Bauman 1991b, 5)

It took at least two centuries of theoretical rivalry and practical failure for the technoscientific community to recognize that what was assumed to be true yesterday may turn out to be false today. (This remains the case regardless of whether one uses a Whiggish view of the cumulative success of science, as seen in Butterfield 1965, or the paradigmatic shifts described in Kuhn 1972.) It also took some time for them to realize that there can be more than one interpretation of the collected data (a fact which was eventually to be treated as the problem of incommensurability). And here is a point where the behavior of the technoscientific community is more attractive than that of the traditional (e.g., Catholic) church: Whereas traditional religious hierarchies relegate exclusively to themselves authority regarding the truth, in the technoscientific community competing interpretations can see the light of day and may even overthrow an established, authoritative model or view.

Two related consequences of the prominence of technoscience have come to light by the end of the twentieth century: First, there is an ongoing desire to uphold an ontological foundation, supported by epistemological scaffolding, that will provide social cohesion and political stability. Second, to maintain a formal commitment to a foundation may require an ongoing adherence to the consideration and testing of alternative views and models that may alter the very foundation upon which the technoscientific worldview is formulated. These consequences give rise to a recognition of the need to remain critical in the self-evaluation of the views and principles of the accepted worldview.

If technoscience can afford humanity an adequate foundation, then and only then can it successfully compete with religious doctrines. Once the public has been promised an adequate replacement, it will expect the technoscientific community to deliver on this promise. It is unclear whether the technoscientific community itself ever made such a promise explicitly to the public or the public assumed this to be the case. Regardless of the exact chain of events that may have led to this situation, it is fair to say that by the end of the twentieth century, and especially with the proliferation of the many innovations whose fruits have been enjoyed across the Western world, public perception of technoscience is such that the public perhaps expects more than the technoscientific community can comfortably provide.

What is at issue is not the privilege of technoscience or specific technoscientific knowledge claims, as some sociologists maintain; rather, what is at issue is the public's expectation that technoscience will provide a set of answers on which it can solidly rely. This expectation leads to a definition of technoscience in terms of what people would like it to be and not in terms of what it is or can be. One issue for consideration, especially in light of the critiques of technoscience in the twentieth century, from the Vienna Circle to the social constructivists and the feminists, is that of the ambiguity of technoscientific data. Would technoscientists themselves make claims for the ambiguity or for the certainty of their data?

From Ambiguity to Anxiety

Some amount of tolerance and generosity would allow that any pronouncement is inherently ambiguous, for no pronouncement can express all the conditions under which it was formulated nor all the conditions under which its truth value remains warranted. As such, any pronouncement is open to interpretation and therefore is inevitably rife with ambiguities. We learn to tolerate ambiguities because we know under what circumstances their range can be accounted for. We also permit some level of generosity for the readers or judges of the said pronouncements, so that implicit assumptions can be made explicit. Within this context, pronouncements, that is, any set of technoscientific statements, are open for testing.

The technoscientific enterprise is committed to minimizing ambiguities so as to provide a better approximation of the truth about reality, the universe, or the world as we know it. Even when ambiguities in one level seem to be resolved, this does not guarantee that additional ambiguities will not arise at a different level. But this very process is taken for granted by technoscientists, and the most they may be able to accomplish is "higher" levels of ambiguities so that "lower" levels become somehow more certain.

For example, Popper's concern with testability and falsification is a concern not with truth but with an approximation of the truth (1959, 1963). The technoscientific community, with its empirical and rational arsenal, has relinquished the goal of a full comprehension of the truth for a different one. The goal, as Popper and the Vienna Circle understood it, is to know for sure enough things about nature that we can negotiate our survival. When we know for sure that a hypothesis is false, we know quite a bit; we may postulate auxiliary hypotheses, revise our original hypothesis, and in the process learn how to avoid errors and venture into territories that may provide hints of what we may eventually come to know.

The process in which the technoscientific community engages may lead to anxiety: Every principle, model, theory, or hypothesis one formulates is likely to face rigorous testing that may prove it false. Even when one sits, as it were, on the shoulders of technoscientific giants, one is not guaranteed success (in the sense of achieving a closer approximation to the truth). To some extent, this condition is part of the technoscientific gestalt. What makes this gestalt unique is that its provisionality is by design and not by default and, furthermore, that those joining this group activity are conscious of this framework before they join it. Moritz Schlick conceded this point in 1925, along Cartesian lines, making explicit references to the "abyss" of doubt mentioned earlier. It is worth quoting at length his confession:

> When we stand with such thoughts on the highest peaks of skepticism, a shutter of intellectual anxiety comes over us. We are seized with dizziness, for we glimpse an abyss that seems bottomless. This is a point at

which the paths of the theory of knowledge, of psychology and—as I hope I may add—of metaphysics intersect and suddenly break off. We cannot be satisfied, once we have looked into the abyss of doubt and uncertainty and have drawn back from the brink, merely to return unmoved to the land of common sense. We cannot comfort ourselves with the thought that such doubts are fruitless and that despite them the sciences enjoy a firmly grounded existence. We do not want to ascend once more into the light of science until we have taken full measure of the depths of the knowing consciousness. Epistemology is not as fortunately situated as the individual sciences, which can leave the verification of their foundations to a more general discipline; the theory of knowledge is concerned precisely with the ultimate presuppositions of *all* certainty. We can hope to overcome universal doubt only if we strip the difficulty of its wrappings and face it calmly. (Schlick 1985, 118)

So, one may ask, why pursue the activities and hard work of technoscience in the face of this anxiety? Can the community that is devoted to a calm engagement with the complexities of reality ever accomplish its tasks and transcend the conditions and consequences of anxiety?

According to Stephan and Levin, there are three reasons that motivate (techno)scientists: curiosity, professional status, and financial rewards, that is, "puzzle, ribbon, and gold" (1992, 8). As already noted by Kuhn, the mainstay of "normal science" is puzzle solving, as if all participants are initiated into a game (i.e., a paradigm) so as to put together as many pieces of the puzzle as they can master (Stephan and Levin 1992, 18). Following Merton, Stephan and Levin claim that an additional motivation for doing science is peer and public recognition. Whether recognition is understood as professional reputation or institutional status, technoscientists are concerned with and respond to rewards emanating from their professional activities (Stephan and Levin 1992, 18–20). The third factor that motivates technoscientists is monetary reward, that is, rewards that come in the guise of research grants or that are directly given as prize money or salary increases (Stephan and Levin 1992, 20–22).

If indeed the three main factors that motivate technoscientists are "puzzle, ribbon, and gold," then it is quite clear why anxiety inevitably creeps into their minds. What if the puzzle cannot be solved quickly, if at all? What if the puzzle is solved but there is no recognition of the solution? What if multiple solutions are offered simultaneously? What if someone steals the solution? What if funding agencies are strapped for funds and no grants are awarded in a particular field of research? These are haunting questions for anyone being initiated into the technoscientific community. It is not surprising that anxiety would run high among technoscientists faced with such questions after spending years in training and then having to keep up with the latest global developments in the field.

One way in which Stephan and Levin document the creeping anxiety of technoscientists, and even a way of explaining certain demographic realities of their community, is to focus on age. They conclude that in general younger scientists are more productive than older ones and that "there is a strong relationship between age and the ability to do path-breaking work" (Stephan and Levin 1992, 156). In the present context, the focus on age, and especially on the premium given to youth, produces a progressively more pronounced sense of anxiety within the technoscientific community. Every individual within this community is bound to do less work and less-important work as the years go by. Instead of accumulating knowledge and expertise, one loses touch with the most exciting and demanding developments in one's own field of research.

As Stephan and Levin argue, there is another important factor that dominates the technoscientific community and that also contributes to a sense of anxiety. They call it "RPRT: being at the right place at the right time" (Stephan and Levin 1992, 158). If we recall the Stoics and their demarcation between those things that are within human control and those that are not, RPRT obviously falls into the category of those things over which humans have no control. If one cannot control the circumstances that will dictate one's success or failure, then it is obvious that hard work and dedication alone will not assure an anxiety-free working environment.

Stephan and Levin also found that the increased pool of trained technoscientists and the increase in their funding requests, coupled with a decrease in available funding, bring about increased competition. Increased competition has at least three detectable consequences: paper inflation, aversion to risk taking in research, and misconduct. Paper inflation is shorthand for the substitution of quantity for quality, since there is increased pressure to publish as much as possible in the shortest amount of time. Second, aversion to risk taking is clearly understood as a strategy to ensure continued funding for research programs whose partial success has been fully demonstrated. In a climate of fierce competition, funding agencies and institutions will refuse to gamble on a project that may yield no results whatsoever. And finally, misconduct is noticeable when researchers take shortcuts in order to be "productive" and "efficient" in a competitive atmosphere. What is important is to deliver the expected results, no matter how and no matter how accurate they turn out to be (Stephan and Levin 1992, 162–164).

These three sets of consequences of the technoscientific community's competitive environment have an adverse impact on the community's long-term progress. Instead of allowing ambiguity to play its role in the technoscientific life of researchers, the play of ambiguity—with its promises of new interpretations of old materials—is replaced with the horror of anxiety. Anxious practitioners are under tremendous pressure, some of which is not of their own doing and most of which is unproductive. Instead of motivating them, providing more incentives to work harder, the competitive pressure described by Stephan and Levin mystifies and terrorizes; it may even paralyze.

Stephan and Levin make clear in their study of the workings of the technoscientific community that the epistemological issues discussed above only set the stage for a more acutely felt anxiety among practitioners of the trade. What may seem to be philosophical musings concerning ontological and epistemological issues are instead concerns reflected in the structural organization of the field, influencing the actual behavior of technoscientists. One may argue that the situation of technoscientists differs little from that of any other group of people working in the late twentieth century: The human condition rests on the brink of the abyss, with its existential futility, absurdity, and ennui.

Prior to World War II (as I discussed in Chapter 2), Freud (1989) wrote about the similarities he found between the developmental stages of individual psyches and the process of civilization in general. The line of argument he develops in this text states that just as the child moves from the gratification of the "pleasure principle" to the restraints of the "reality principle," so does civilization move from the former to the latter. Though using a different terminology, Freud echoes Rousseau's notion of the shift from the "noble savage" to the civilized citizen. As Freud says: "What we call our civilization is largely responsible for our misery . . . we should be much happier if we gave it up and returned to primitive conditions" (Freud 1989, 38).

Whereas Rousseau emphasizes the sense of self-preservation and

pity of our ancestors (Rousseau 1964, 95, 130–131), Freud emphasizes sexual drives: "Present-day civilization . . . does not like sexuality
as a source of pleasure in its own right and is only prepared to tolerate
it because there is so far no substitute for it as a means of propagating
the human race" (Freud 1989, 60). When sexuality is "tolerated" and
not celebrated, humans experience frustration, perhaps even confusion concerning a "natural" instinctual drive that has become socially
suspect. Why would humans accept this situation? According to
Freud, a bit of security and anxiety reduction is worth the price of losing a bit of happiness (Freud 1989, 73).

If we have encountered before the tendency of religious doctrines
and technoscientific models to provide cognitive stability, or "security" in Freud's sense, then it makes sense that the cultural parameters
informing the establishment and development of civilization (i.e.,
Western, Eurocentric civilization) would ensure stability and security
as well. Freud called this security an "agency" that is "like a garrison
in a conquered city" (Freud 1989, 84). The "agency" is what he defines as the alter ego or superego, an influential part of the human
psyche whereby social norms and conventions are internalized and become part and parcel of the way every individual perceives the world
and her or his own ego formation and behavior. If we have observed
earlier how ambivalence and ambiguity lead to cognitive anxiety, now
we are introduced to a parallel formulation: how the mismatch between one's instinctive behavior and the social context within which it
is manifest turns into the anxiety of not being able to react and respond
outside of a prescribed set of rules and regulations. As Freud says:

> If civilization is a necessary course of development from the family to
> humanity as a whole, then—as a result of the inborn conflict arising
> from ambivalence, of the eternal struggle between the trends of love and
> death—there is inextricably bound up with it an increase of the sense
> of guilt, which will perhaps reach heights that the individual finds hard
> to tolerate. (Freud 1989, 96)

"Guilt" is another term one can use to describe the kind of anxiety
that torments humans. This does not mean that guilt and anxiety are
identical or that one adequately describes the other; all it means is that
the one is analogous to the other in the sort of consequences that
come about when it is felt and acted upon. This guilt—or anxiety—
leads Freud to speak of civilization as being "neurotic" in some gen-

eral sense (Freud 1989, 110). As we have seen in Chapter 2, Freud
helps bring together the concepts of guilt and anxiety as well as the
concepts of the technoscientific enterprise and development of civilization in its quest for control. If civilization is liable to itself in the
threat it feels from its own development, then "unhappiness" and a
"mood of anxiety" are bound to set in (Freud 1989, 112).

What brings about a pervasive sense of anxiety in modern society is
not limited to the conflict between the individual and the society
within which she or he interacts with others; rather, there was a pre-
monition before the onslaught of World War II (but probably with a
keen eye on the results of World War I) concerning the power to exterminate all humans. The conquest of nature, as imagined for centuries
by the leaders of scientific revolutions, has turned into the nightmare
of the annihilation of humanity. What is at stake is not only the capa-
bility to destroy others but the conscious realization that this capabil-
ity exists.

For Freud, the conscious realization of the reality principle means
acknowledging not simply a set of obstacles one may or may not overcome but also a set of destructive options one may pursue. This self-
conscious position of the psyche causes a great deal of personal anxiety, an anxiety that cannot be laid to rest since no external assurance
will ever terminate internal conflict. Though coming to the same issue
from a sociological perspective, Bauman finds himself arguing about
the "privatization of ambivalence," that the impossibility of "eradi-
cating" ambivalence has become a "personal affair." As such, ordering the universe is "an individual task and personal responsibility."
Just as the effort is personal, any failure to make sense of the universe
is perceived as a personal failure. The blame for the "failure" brings
about "the feeling of guilt" (Bauman 1991b, 197).

Though not as concerned with psychology and psychoanalysis as
Freud, Bauman too connects modernist anxiety with personal guilt
feelings. But unlike Freud, who seems to resign himself to the fact that
humans may be condemned to live under conditions of anxiety, guilt,
and neurosis, Bauman critically examines ways in which human anxiety may be mediated, if not eliminated. The "expert" is singled out as
a "translator" who can simultaneously deal with the subjective anxieties of individuals and the objective knowledge claims forwarded by
the technoscientific community. The expert shuttles between these two
spheres of conduct, delivering messages from one group to another,

while maintaining the goodwill of both. Eventually, the work of experts seems necessary so that widespread breakdowns will not devastate civilization (Bauman 1991b, 199–230).

Experts, as I have argued elsewhere (Sassower 1993), are bound to fail even when they do their best because the conditions under which they operate make it impossible for them to ever fully succeed. These conditions are both cognitive and emotional; they are epistemological and psychological. These conditions have a tradition of their own that makes it difficult to change them quickly, if they can be changed at all. They have an inertia that takes over our worldview or that informs the worldview we have learned to accept.

And what would count as success? If experts convince individuals that a technoscientific model is true, they know right away that three issues will eventually surface.

First, any truth claim is bound to be challenged and may turn out to be false, so to convince anyone of the truth of anything may verge on dishonesty. Second, if the provisional nature of a truth claim is not made explicit and if indeed the public is convinced by a truth claim, then the expert differs little from the priest who convinces the congregation of the truth of church doctrines. Thus, there is an inherent danger of indoctrination or plain dogmatism. And third, even if the truth claim is confirmed time and again, there is also the danger that since the public does not have a mastery of the technical vocabulary of the technoscientific community, the public's reliance on expert translation is tantamount to its having faith in the expert. Once again, are we not back in the religious fold?

One could come up with additional complications concerning the feedback loop established between the public, the technoscientific community, and the expert. For example, what if the technoscientific community internalizes the fears and anxieties of the public and thereby caters to these fears and anxieties? What would be the nature of the technoscientific enterprise if it were to constantly pander to the public and its expectations? Would we not have a more conservative, less imaginative, and less risk-taking enterprise? Would we not have a less critical and less honest enterprise because any mistake, error, or failure would be covered up? Instead of having debates open to the public so as to inform them of the ambiguities of technoscience, there would be debates about what to disclose and what to hide from public scrutiny.

In the face of incessant anxiety, a condition that seems beyond re-proach or repair, there slowly creeps in an admission that nothing can be done to change the situation. If one cares, as for example Pierre Duhem cared, about this situation, some deep pain begins to manifest itself. This brings us to anguish.

From Anxiety to Anguish

In 1908 Duhem announced (in conjunction with a quotation from Theon) "the impossibility of astronomy's ever discovering the *true* hypothesis, the one which conforms to the nature of things" (Duhem 1969, 9). The quandary facing physicists throughout history, then, is the following: "[Astronomy] shows us that man's knowledge is lim-ited and relative, that human science cannot vie with divine science" (Duhem 1969, 21). Whether or not one wishes to follow Duhem's own recommendation "to save the phenomena," it becomes clear that humans remain hopelessly powerless to come up with answers whose efficacy transcends the relative historical positions they inevitably oc-cupy. This is not only frustrating (provoking anxiety) but also painful (provoking anguish).

There is a certain amount of sadness that accompanies the apprecia-tion of one's humble station in life. The greatest intellectual feats may yield no more than the picturesque musings of myth makers and magi-cians. Humans remain in awe before the vastness of the universe and in the face of multiple visions. They proceed *as if* they know some-thing true and believable, as if they have a handle on reality. But at best they come in contact with the fringes of reality, with its surfaces and edges. So they in fact proceed *despite* what they know and do not know, doing the best they can to keep the ball rolling, to keep the process going. The predicament of technoscientists is reminiscent of Albert Camus's recounting of the myth of Sisyphus:

> The GODS had condemned Sisyphus to ceaselessly rolling a rock to the top of a mountain, whence the stone would fall back of its own weight. They had thought with some reason that there is no more dreadful punishment than futile and hopeless labor. (Camus 1991, 119)

Camus explains that Sisyphus is an absurd hero because he is con-scious of his predicament. What both makes him absurd and puts him in a bind is the very possibility we all imagine: that he would be able to accomplish his impossible task. It is possible to imagine someone rolling a rock all the way to the top! But poor Sisyphus cannot and does not. Is his condition any different from the one described by Schlick, of the scientist standing at the brink of the abyss, staring at it, and avoiding his inevitable fall? Yet just as Schlick offers a way out, so does Camus emphasize the notion of hope amid the existential angst.

If this picture makes even partial sense, technoscientists are bound by the existential framework about which Søren Kierkegaard, Camus, Franz Kafka, and Jean-Paul Sartre all wrote long treatises, novels, and plays. These writers all are painfully aware of the human condition, and they frame it in a manner that induces us to realize our personal involvement and potential to overcome our condition, perhaps not in any global, universal, absolute fashion but, rather, in a personal, subjective, and relative way. It is the contention of this chapter, central to the argument presented in this book, that technoscientists are no different from other intellectuals, academics, poets, and artists in their deeply felt distress over the human condition and the future of humanity as we know it.

The case of J. Robert Oppenheimer, the "father" of the atomic bomb, comes to mind when speaking of existential angst and the dread of a reflective thinker about the fate of humanity. Oppenheimer was the director of the Manhattan Project at Los Alamos, New Mexico, during World War II, a project dedicated to the development of a nuclear bomb in the face of the threat of Nazi Germany. That the atomic bomb was developed in record time and that it was dropped on Hiroshima and Nagasaki in 1945 are two known facts. It is also known that after the war Oppenheimer was reluctant to pursue the development of the hydrogen bomb and that he eventually lost his security clearance because of his outspoken public condemnation of the development of any destructive weapons. He proposed international disarmament talks and controls so as to ensure world peace.

In what follows I will not rehearse the mountains of documents and written reports concerning his views and actions. Instead, I will quote some of his statements so as to illustrate his state of mind once the consequences of the use of the atomic bomb became clear. Alice Kimball Smith and Charles Weiner, who edited some of Oppenheimer's letters and recollections, note that after he received a letter from President Franklin Roosevelt and had responded to it, he was highly distressed:

> As Oppenheimer turned felicitous phrases for the President and Tolman, he was undergoing a real crisis of self-confidence. The heady experience of creating a new laboratory and pulling together the disparate parts of the scientific work had been stimulating and euphoric. Then came the reaction. On several occasions in the early summer of 1943 Bacher found Oppenheimer depressed by the magnitude and complexity of the director's task. He told Bacher that he could not go through with it, but Bacher's advice was simple: Oppenheimer had no alternative, for no one else could do the job. (Smith and Weiner 1980, 261)

Though Oppenheimer was a scientist, he found himself more often than he could have imagined in a political role or position; in this context, note an interesting letter to Ernest O. Lawrence of August 30, 1945, concerning the precedent that political decisions have over the concerns of the scientists at Los Alamos (Smith and Weiner 1980, 300–302).

October 16, 1945, was Oppenheimer's last day as the director of the Los Alamos National Laboratory; he wrote:

> If atomic bombs are to be added as new weapons to the arsenals of a warring world, or to the arsenals of nations preparing for war, then the time will come when mankind will curse the names of Los Alamos and Hiroshima. (Smith and Weiner 1980, 310–311)

Leo Szilard was able to receive the support of James Franck, since both worked in Chicago at the time, and the Franck Report against the use of the atomic bomb on Japan was delivered to the secretary of war and the Scientific Advisory Panel, of which Oppenheimer was a member. According to Barton Bernstein,

> Oppenheimer strongly opposed Szilard's petition [drafted after the Franck Report failed to mobilize political reconsideration of the use of the atomic bomb in the war with Japan]. Dropping the bomb was necessary, Oppenheimer told Teller, the best minds were already giving advice, Szilard should not interfere, and furthermore, scientists should not oppose the will of elected policymakers or claim any special responsibility for directing the use of the weapon. (Quoted in Hawkins 1983, xxxvi)

As we shall see in later chapters, there is a continuum of views concerning the different levels of responsibility assignable to scientists. While some claim that once the scientific work has been completed, it becomes the responsibility of politicians, others argue that scientists ought to remain involved and liberally offer their advice (and thereby act responsibly). The case of Oppenheimer is of particular interest because it seems that he shifted his own view from one extreme of the continuum to the other.

In his speech to the Association of Los Alamos Scientists on November 2, 1945, a few months after the war and after observing the results of the Manhattan Project, Oppenheimer said:

> I think there are issues which are quite simple and quite deep, and which involve us as a group of scientists—involve us more, perhaps than any other group in the world. I think that it can only help to look a little at what our situation is—at what has happened to us—and that this must give us some honesty, some insight, which will be a source of strength in what may be the not-too-easy days ahead. (Smith and Weiner 1980, 315)

Though obviously concerned with the fate of humanity in the face of the devastating experiences of Hiroshima and Nagasaki, Oppenheimer was also concerned with the fate of science itself. He compared his own revolutionary era with "the days of the renaissance, . . . when the threat that science offered was felt so deeply throughout the Christian world" and with "the days in the last century when the theories of evolution seemed a threat to the values by which men lived." His comparison led him to proclaim that his own age "has in common with the early days of physical science the fact that the very existence of science is threatened, and its value is threatened" (Smith and Weiner 1980, 316).

Oppenheimer's dramatic comparison is noteworthy because it highlights the deeply felt anxiety that surrounded the scientific community in general and the community of scientists at Los Alamos in particular. If indeed the impact of the development and use of the atomic bomb has become a political instrument of global warfare and if the scientific community has to bear some responsibility for their scientific work, then the pristine environment in which they work, a sanctuary of sorts, disappears. But why was there a threat to science at this particular moment different from any other time in which scientific work was used for destructive purposes? Oppenheimer outlined three reasons:

> One is the extraordinary speed with which things which were right on the frontier of science were translated into terms where they affected many living people, and potentially all people. Another is the fact, quite accidental in many ways, and connected with speed, that scientists themselves played such a large part, not merely in providing the foundation for atomic weapons, but actually making them. In this we are cer-

tainly closer to it than any other group. The third is that the thing we made—partly because of the technical nature of the problem, partly because we worked hard, partly because we had good breaks—really arrived in the world with such a shattering reality and suddenness that there was no opportunity for the edges to be worn off. (Smith and Weiner 1980, 316)

The threat of Nazi Germany during World War II and the fact that many of the scientists working on the bomb were Jewish refugees from Germany and other occupied European nations propelled the scientific community to a level of collaboration never before seen in the history of science and technology. Without analyzing the socio-psychological conditions and tensions under which his leadership was undertaken and without tracing his personal background and the details of his life at the time (something done by the FBI and CIA), let me suggest that Oppenheimer stands as a fascinating and admirable figure in that history because of the few but profound philosophical reflections he offered.

In particular, I would like to continue a brief analysis of his speech to the Los Alamos scientists, because in it he explains the various jus-tifications or reasons that motivated his collaborators and juxtaposes them against his formulation of the scientific ethos. In short, Oppen-heimer demonstrates one of the themes of this book, namely, that technoscientists are neither dupes nor villains but, in fact, sincere re-searchers full of anxiety who are seduced by the scientific ethos to work hard (with an implicit promise of self-policing) for lofty goals that are then subverted (by their own leadership and by military and political leaders) without consultation.

Oppenheimer lists six main reasons why different scientists joined the Manhattan Project: first, the fear that the atomic bomb might be developed by the enemy; second, the feeling that without the bomb it might be impossible to win the war; third, curiosity; fourth, a sense of adventure; fifth, the desire, since atomic weapons were possible in principle, to see what could be done in practice; and sixth, the hope that atomic weapons could lead to a reasonable solution (in American hands) to current and future global conflicts. Oppenheimer believed that any of these reasons might have been good enough to explain the participation of so many scientists in this effort. He went on to say:

But when you come right down to it the reason that we did this job is because it was an organic necessity. If you are a scientist you cannot stop such a thing. If you are a scientist you believe that it is good to find out how the world works; that it is good to find out what the realities are; that it is good to turn over to mankind at large the greatest possible power to control the world and to deal with it according to its lights and its values. (Smith and Weiner 1980, 317)

Oppenheimer's sincerity reveals itself in the expression "an organic necessity," as if this phrase alone can explain everything that every technoscientist feels when asked to embark on a technoscientific project. The "organic" nature of the endeavor is explicable in terms of the ongoing quest for knowledge (and control) of the world and its realities. The sense of "necessity" is linked to a certain appreciation of the inevitability of the technoscientific process, a sense that once humans explore something in the world they will inevitably explore everything in the world. However problematic this phrase may be, Oppenheimer goes on to explain it in terms of the scientific ethos, in a language that is reminiscent of Merton's ideals of science:

It is not possible to be a scientist unless you believe that it is good to learn. It is not good to be a scientist, and it is not possible, unless you think that it is of the highest value to share your knowledge, to share it with everyone who is interested. It is not possible to be a scientist unless you believe that the knowledge of the world, and the power which it gives, is a thing which is of intrinsic value to humanity, and that you are using it to help in the spread of knowledge, and are willing to take the consequences. And, therefore, I think that . . . anything which is an attempt to treat science of the future as though it were rather a dangerous thing, a thing that must be watched and managed, is resisted not because of its inconvenience . . . but resisted because it is based on a philosophy incompatible with that by which we live, and have learned to live in the past. (Smith and Weiner 1980, 317–318)

Oppenheimer's complaint at the end here refers to General Leslie Groves's concern with the secrecy of the project, and in particular to his method of compartmentalization (which will be discussed further below).

Oppenheimer ends his speech with four specific recommendations according to which scientists ought to fulfill their obligation to their community and to society at large. First, he advises scientists to be careful and honest about their knowledge and about the way they can answer technical questions. Second, he advocates fraternity among scientists within nation-states and across borders, so as not to lose confidence among fellow scientists. Third, he urges scientists not to lose faith in the value of science and its ability to do good in the world. And fourth, he reminds scientists of their bonds with their fellow humans and that scientists have duties that go beyond their professional activities (Smith and Weiner 1980, 324–325).

Heinar Kipphardt, in his play about the hearings on Oppenheimer's security clearance, culls from the three thousand typewritten pages of the hearings some interesting reflections by one of the members of the Personnel Security Board, Ward V. Evans:

> I cannot reconcile these interrogations [of Oppenheimer's private life and his alleged disloyalty because of his associations with Communists] with my idea of science . . . On the other hand, it was the physicists themselves who started the whole thing when they turned their profession into a military discipline . . . perhaps science, too, must bow to the absolute claims of the state. Now that science has become so important. (Kipphardt 1968, 25)

Evans may be representing the view that the status traditionally accorded to the scientific community may be changing, that it no longer can claim objectivity and neutrality within the culture that supports it. Assuming that political objectives represent the will of the people (at least of a majority) and assuming that the military abides by the wishes of political leaders, then turning science into a military discipline is the wish of the people. If this is the case, any claim by the scientific community to disregard political and military pressures seems unreasonable. Evans continues to say:

> I can see two kinds of development. The one is our increasing control over nature, our planet, other planets. The other is the state's increasing control over us, demanding our conformity . . . But how can a thought be new, and at the same time conform? (Kipphardt 1968, 25)

Evans clearly appreciates the predicament of the scientific community: Though they invent instruments for one purpose, these same instruments may be eventually used in ways unforeseen by their originators; the television monitor, for example, used as a means of surveillance. The question Evans asks is echoed by Oppenheimer when he states that "you cannot produce an atomic bomb with irreproachable, that is, conformist ideas. Yes-men are convenient, but ineffectual" (Kipphardt 1968, 40–41). Whereas security officers are concerned that there is too little bureaucratic conformity, scientific leaders like Oppenheimer are concerned that there is too much. Secrecy may be crucial for a war effort, but it is detrimental in trying to solve techno-scientific problems.

Among the witnesses called in by the board was Edward Teller, a participant in the Manhattan Project and the champion of the hydrogen bomb. To some extent, Teller expresses the optimistic view of many of his colleagues when he describes scientific discoveries as neutral undertakings whose eventual use benefits humanity:

> Discoveries in themselves are neither good nor evil, neither moral nor immoral, but merely factual. They can be used or misused. This applied to the internal-combustion engine, and it applies to nuclear energy. By painful experience, man has always learned in the end how best to use them. (Kipphardt 1968, 94)

Nuel Pharr Davis, in his book on Lawrence and Oppenheimer, explains the precarious position into which the community of physicists were put once Szilard's letter was signed by Einstein and forwarded to Roosevelt (1968, 99). Davis brings up Compton's assertion that it would be "better to be a slave under the Nazi hell than to draw down the final curtain on humanity" (Davis 1968, 131). Davis also quotes Oppenheimer's words in regard to the concerns about the completion of the Manhattan Project:

> Looking ahead into 1945, Oppenheimer could see a date when war deaths would become his responsibility. "We had the feeling—perhaps wrong—that a hundred thousand lives, maybe more, might depend on whether this thing was ready August 1, September 1, or October 1," he said. (Davis 1968, 219)

Davis quotes Oppenheimer ten years after the end of the war: "I find myself profoundly in anguish over the fact that no ethical discussion of any weight or nobility has been addressed to the problem of atomic weapons" (Davis 1968, 329–330). Oppenheimer's feelings of ambiguity and anxiety are being compounded with a strong feeling of anguish, "despair," as Gerald Holton calls it. Davis continues to quote Oppenheimer:

> What are we to make of a civilization which has always had in it an articulate, deep, fervent conviction, never perhaps held by the majority, but never absent: a dedication to *Ahimsa*, the Sanskrit word that means doing no harm or hurt, which you find in Jesus and simply and clearly in Socrates, what are we to think of such a civilization, which has not been able to talk about the prospect of killing almost everybody, except in prudential and game-theoretic terms? (Davis 1968, 329–330)

Ferenc Morton Szasz explains that the Los Alamos site was primarily established for the sake of efficiency and secrecy, especially since General Groves insisted on compartmentalization. He says: "After the war, Leo Szilard complained bitterly that Groves's insistence on Compartmentalization actually hindered the development of the bomb by eighteen months. Groves, however, always defended his position" (Szasz 1984, 16). Szasz gives a good summary of the two major positions regarding the necessity of dropping the bombs on Japanese targets, the one claiming that dropping the bombs would end a costly war, and the other claiming that Japan was already on the verge of collapse without this particular threat (Szasz 1987, 149–152). He quotes Oppenheimer as saying that "the decision [to use the bombs against an enemy] was implicit in the project" (Szasz 1984, 152). This is an important point for my discussion of the predicament facing technoscientists enlisted to develop a new technology. As he writes:

> After the Trinity success, the only action that Truman might have taken would have been to *stop* the process. But this was virtually impossible. By August of 1945, the furies of history, wrapped in the garb of the Manhattan Project, had assumed a momentum all their own. (Szasz 1984, 156)

He does admit that President Harry Truman had the following entry in his diary:

> I hope for some sort of peace—but I fear that machines are ahead of morals by some centuries and when morals catch up, perhaps there'll be no reason for any of it. I hope not. But we are only termites on a planet and maybe when we bore too deeply into the planet there'll [be] a reckoning—who knows? (Quoted in Szasz 1984, 157)

Despite Oppenheimer's public confessions concerning the anguish and guilt he felt, Szasz admits:

> Yet not all the Los Alamos scientists felt the same anguish and guilt. As Freeman Dyson has noted, many of the Cornell physics department resented Oppenheimer's confession. They considered themselves no more guilty than anyone else who had made lethal weapons during World War II. Those who remained at Los Alamos, under its new director, Norris Bradbury, felt the same way. After the failure of all international plans of control, combined with Soviet advances in weapons, this position gained in popularity. (Szasz 1984, 174)

In the next chapter I wish to explore in more detail the relationship between the ambiguity, anxiety, and anguish endemic to the technoscientific community and the views on responsibility adopted by its members. My concern is not to argue that the members of this community ought to have a sense of responsibility; rather, my concern is with the particular expressions of their sense of responsibility. My claim is that since the stakes associated with technoscience in general and with the development of technoscientific tools during World War II in particular are and were so high, there is no way to overlook, avoid, or minimize the question of responsibility.

Yet to suggest that all members of the technoscientific community should translate their ambiguity, anxiety, and anguish into proactive expressions of their personal conviction and responsibility is too broad and too vague. Would it suffice if they declared their remorse after the fact? Would it help if they made their debates public? Would it reassure the public if they lamented their work and portrayed their anguish as an insurmountable objection to further research and development? Should they consult the media as much as their colleagues? Using the records of the personal experiences of some of the members of the Manhattan Project, I try to suggest to what extent a postmodern twist to the question of responsibility may be helpful in the future.

4

The Postmodern Option:
A Dialectical Critique

In this chapter I examine the views expressed by the leaders and members of the Manhattan Project concerning their personal responsibility for the development of weaponry capable of the mass destruction of human lives. In order to articulate these concerns I will probe three sets of issues: some of the material that frames the technoscientific contribution to the cruelty and destruction of World War II, the problems with generalizing from this focus, and the sense of responsibility that emanates from these events. My focus on technoscience overlooks many other factors that may be thought more fundamental in assessing these events. Yet my own focus may enrich the discussion of personal responsibility in the face of political and military decisions. It remains an open question whether a postmodern orientation (however flawed and unsatisfactory) enhances the sense of responsibility rather than dispelling it. At the end of the chapter I will recount some literary and archival materials that illustrate the deep ambiguity, anxiety, and anguish felt by members of the technoscientific enterprise.

The Centrality of Technoscience in World War II

For the sake of brevity, I shall follow generations of critics and use the labels "Auschwitz" and "Hiroshima" to embody the concerns with a cruelty and destruction never previously attempted on such a large scale. Theodor Adorno's words concerning the unique nature of Auschwitz still ring true today:

After Auschwitz, our feelings resist any claim of the positivity of existence as sanctimonious, as wronging the victims; they balk at squeezing any kind of sense, however bleached, out of the victims' fate. And these feelings do have an objective side after events that make a mockery of the construction of immanence as endowed with a meaning radiated by an affirmatively posited transcendence. (Adorno 1973, 361)

Auschwitz was a turning point in our thinking about wars and about civilian fatalities; the deliberate murder of millions of civilians, not as a side effect of urban bombing but as a separate military agenda, has never been so ruthlessly pursued and at such an enormous level. The phrase "after Auschwitz" has since denoted this cultural turning point. Moreover, World War II is still *our* war, even if we were too young or too far away to experience it firsthand: It is the war of our parents, the war that defined their lives and therefore our lives, too. When our parents die, this war will recede into the dark recesses of the archives, into books and libraries, into museums. Moreover, this is our war because it was the first war to be technically delivered across the globe with a media blitz that had an unprecedented cultural impact. It is our war because of the extensiveness of its reports and documentation, which infiltrate the remotest and most secluded segments of the population.

Perhaps it is also our war because of the centrality that technoscience played in it. This is not to say that previous wars had not benefited from the services of scientists, engineers, and technicians; rather, what is shocking is not only that such a large number of technoscientists served military and political leaders but that they almost dictated the rules under which World War II was fought. Alan Irwin concedes that most organizations involved in the integration of science and social policy, those who are concerned to bring about a process in which citizens are intimately engaged in deliberating over and implementing scientific and technological loans and products, share

a fundamental belief in the centrality of scientific development to the future of society—and a belief (whether as part of a social democratic or more vaguely liberal ideology) that a better informed citizenry can play a crucial (but essentially reactive) role in this development. The future should indeed belong to science. (Irwin 1995, 14)

The "science-centered" worldview espoused in his treatise leads Irwin to argue that there is some truth in the belief that because of the centrality of science in the modern world and the general public's basic ignorance of its principles and workings, *"greater scientific under-standing* amongst the public will lead to *greater acceptance and sup-port* for science and technology" (Irwin 1995, 26). Influenced by the sociology of scientific knowledge, Irwin maintains that the integration between the needs of science and society can be successful if "the question . . . becomes not *whether* science should be applied to envi-ronmental (and, of course, other) questions but rather *which form* of science is most appropriate and in *what relationship* to other forms of knowledge and understanding" (Irwin 1995, 170).

It seems that every generation of intellectuals of whatever philo-sophical orientation must account for Auschwitz and Hiroshima not only as instances of inhuman mass destruction but also as events whose perverse success depended to a large extent on the technoscien-tific community. If these intellectuals do account for these events in particular, then they may account for the interface between nuclear technology and postindustrial democracy, as Joseph Morone and Ed-ward Woodhouse do. Though their historical survey does not deal with Los Alamos as such or with the development of nuclear bombs, they do concede that "nuclear power scores worse on these five crite-ria than virtually any other technology: it is unfamiliar, involuntary, and not under personal control, with low perceived benefits and a dramatic potential for catastrophe" (Morone and Woodhouse 1989, 94). Their analysis goes on to consider public perception and its ef-fects on political decision-making processes and in this fashion en-hances a similar argument about the interface between postmodern-ism and postindustrial democracy in America (Sassower 1995).

After reflecting on the public concern with its control over its im-mediate environment (as we did in Chapter 2) and after considering the shift from ambiguity to anxiety and anguish that members of the technoscientific community feel in light of increased public pressure for the delivery of methods of control (as we did in Chapter 3), it makes sense to turn from the examination of modernity and Enlight-enment ideals to some current forms of postmodernity. When I turn to postmodernism, I am offering a continued critique and not a way to overcome all intellectual and practical obstacles. In this sense, then, the present work is a continuation of my earlier works on expertise (Sassower 1993) and on the cultural context within which techno-scientific projects are undertaken or abandoned (Sassower 1995). I should reiterate that I bring up postmodern ideas and strategies in order to explore their potential and underscore their weaknesses in light of Auschwitz and Hiroshima; these events stand out as crucial cases that must be accounted for; they serve as yardsticks too dangerous to ignore. Put differently, if the postmodern ethos fails to deliver a message of peace and ethical engagement in the face of these two events, then it fails on all other scores as well.

Among the many so-called postmodernists, Lyotard deserves special attention for addressing the intellectual injunction I set up above. First, he is infinitely more sensitive than most in appreciating the complicity of storytellers or of critics who analyze an argument, a narrative, or a theoretical model:

> We are always within opinion, and there is no possible discourse of truth on the situation. And there is no such discourse because one is caught up in a story, and one cannot get out of this story to take up a metalinguistic position from which the whole could be dominated. We are always immanent to stories in the making, even when we are the ones telling the story to the other. (Lyotard and Thébaud 1985, 43)

Lyotard's sense of immanence accounts for Adorno's disillusion with immanence as transcendence within the historical context that he defines as being "after Auschwitz." Being part of the story we tell diffuses the possibility that some truth will be revealed beyond the confines of the story itself. This way of conceiving the situation also accounts for personal experiences in the face of pretense to an objectified truth. And finally, Lyotard's conception accounts for the ethnographic and sociological move toward reflexivity (see, e.g., Bourdieu and Wacquant 1992), at least in the sense that we are complicit in the stories we formulate and judge.

Second, Lyotard is one of a handful of philosophers who constantly remind their readers that their views and linguistic acrobatics must be tested against the reality of the Holocaust. This is true whether one speaks of truth and evidence in the courts of law, as Lyotard does in *Just Gaming* (Lyotard and Thébaud 1985, 74) or whether one addresses the problematics of *The Differend* (1988). For example:

> But, with Auschwitz, something new has happened in history . . . the facts, the testimonies . . . the documents . . . and the names, finally the possibility of various kinds of phrases whose conjunction makes reality, all this has been destroyed as much as possible. (Lyotard 1988, 57)

In short, *to be philosophical is to remain bound by the reality and consequences of Auschwitz.* If the horrors of World War II can escape the framework of one's theory, if their reality can be erased, so to speak, then the theory is bankrupt and needs radical reconstruction to be of any value whatsoever.

My concern with postmodern technoscience is therefore not limited to the problematization of science and technology in the postmodern world nor to useful critiques of science and technology that may bolster the validity and credibility of postmodern principles and theories (see Sassower 1993, 1995). Instead, mine are ethical concerns that parallel those expressed so eloquently by Susan Rubin Suleiman:

> Is it possible to theorize an *ethical* postmodernist subjectivity without recourse to universal values, but also without the innocent thoughtlessness of the "happy cosmopolitan"? Is it possible to argue that such an ethical postmodernist subjectivity has *political* (collective, relating to the public good) import and relevance? Finally, is it possible to argue for a political postmodernist praxis? In plainer words, what do we do if words fail and the shooting starts? (Suleiman 1994, 230)

One may argue that these questions are not new, for they haunted Socrates and Aristotle, Spinoza and Kant; they set in motion the present-day obsession with fulfilling the Enlightenment project, as Jürgen Habermas is fond of doing. Besides, one may agree that these questions are made explicit by the Enlightenment leaders themselves, since their views are not all homogeneous; that is, one is better served characterizing this period in the history of ideas as enlightenments rather than as *the* Enlightenment (for more on this, see Ormiston and Sassower 1989, Ch. 3). If we appreciate the range of perspectives that emanate from this period (perspectives ranging from Hume's to Kant's), we become more sensitive to a dialectical tension between competing epistemological ideals that are deemed worth pursuing in the name of science and for the sake of collective peace and personal happiness.

We admit that we want to have it both ways: We want to legitimate our subjectivity, individuality, personality, ethnicity, and even psychopathology, and at the same time we appeal also to the objective, universal, value-neutral, and equally valid stance of all humans on this earth. We ask others to listen to us, not with generosity and pity, but with the sense that our words ring true for them as well, for the appeal to transcend singularity (in Kant's sense) is exactly what makes our story so interesting, so compelling (or so we have been trained to believe).

Adorno undermines transcendence, and Lyotard injects opinion as a substitute for truth; Suleiman worries about the personal shifting, as

Stuart Hall says (1992, 277), from the autobiographical to the authentic and authoritarian without recourse to any yardstick outside the subjective. Can postmodernism save the day?

I am not a defender of the postmodern faith, for how can anyone defend a faith whose adherents refuse to be faithful to it (see, e.g., Hoesterey 1991)? Nor am I a promoter of technoscience, for its virtues and fruits are universally enjoyed and applauded without the need of the voice of academics. So my concern with the political setting under which a critical evaluation of technoscience in the age of postmodernity must be conducted is a concern with the role of the philosopher as a responsible auditor after World War II, that is, as someone who dares to translate the language game of specialists into the language of ordinary citizens who do not wish to be bothered with the horrors of the past and the nightmares of the future.

Particularity and Universality

Philosophical translation requires at least two ingredients for it to be even partially completed. On the one hand, the translator must be willing to become familiar with multiple literatures. This is difficult because of jargon barriers and the fear of intellectual instability: One is always between too many stools. On the other hand, the translator must be willing to err and to do so publicly, that is, to bear the scorn and humiliation of merciless critics who are all too happy to prove their prowess in their specialization. So self-doubt, the scorn of specialists, and the bewilderment of the general public accompany the work of the philosopher as auditor and translator (for more on this, see Sassower 1995, Ch. 1).

Perhaps the case of accounting for Auschwitz and Hiroshima can illustrate this predicament. What does it mean to say that Auschwitz and Hiroshima stand out as two modernist realities whose uniqueness must be reconciled with their universality? This means in the present context that one walks a tightrope between insisting on the singularity of these realities (in Adorno's sense) and fitting them into a historically explicable whole (in Raul Hilberg's sense, 1973). How can these events be both unusual and typical? This is the dilemma facing anyone attempting to be part of a group and apart from it, trying to hold onto one's tradition and customs while adapting or assimilating to the culture in which one lives.

When Aristotle talks about friendship in the *Nicomachean Ethics*, he talks not about his particular friends but about friends in general; when Donna Haraway (1992) talks about her dogs and cyborgs she describes not simply her own experiences but, rather, experiences everyone has had. From Aristotle to Haraway, the appeal is on behalf of the universal through personal testimonies. The personal is contextualized broadly enough to become more than a subjective account. Similarly, the general concern with the Jewish Holocaust is not simply a perverse or necrophilic fascination with the fate of an unfortunate people or with a sadistic maniac named Adolf Hitler. Rather, it is an attempt to figure out what makes this aberration in the modernist world, this particular experience, a concern of universal proportion.

But was the Holocaust indeed an aberration? Bauman (as already seen in Chapter 1) passionately argues that the Holocaust was not an aberration at all but followed the expected pattern and development of the modern age. As such, it should not be left to specialists and relegated to a historical niche; rather, it requires the rethinking of modern society as we know it and therefore a reformulation of sociology as well. Bauman points to the inherent link between the particular and the universal, whether the link is intended to exemplify, to verify, or to confirm the universal through the particular or intended to problematize, to undermine, or to falsify the universal because of the appearance of the particular. But is the Holocaust not unique, in either Adorno's or Lyotard's sense?

There is a subtle shift from seeing the Holocaust as an aberration, as something whose appearance does not threaten the entire apparatus of modernity, to seeing the Holocaust as a unique occurrence that demands our full attention. What is unique about the Holocaust? In Bauman's view,

> Modern Holocaust is unique in a double sense. *It is unique among other historic cases of genocide because it is modern. And it stands unique against the quotidianity of modern society because it brings together some ordinary factors of modernity which normally are kept apart.* (Bauman 1991a, 94)

Perhaps it is these senses of the uniqueness of the Holocaust that have made it so overwhelming and frightening: It is not unique after all. As a fully modernist event, it does not foreground a single factor, disturbing the entire modernist equation, that one can isolate, focus on, and dispose of once and for all (e.g., a particular leader or ideological pamphlet).

The Holocaust and the atomic bomb, Auschwitz and Hiroshima, are cases that demand attention beyond their immediacy, that haunt the very conception of humanity that philosophers have portrayed over the years. Of course, it would be most convenient to claim that what happened then only happened because of a certain constellation of circumstances that were pertinent to certain historical contingencies from which one cannot generalize (see Dawidowicz 1975). Put differently, it would be convenient to see Auschwitz and Hiroshima as not repeatable (in the sense that experiments that are nonrepeatable are nonscientific) and therefore needing to be named, marked, and shelved in museums. If these events are unique (because unrepeatable), then we should not bother learning about them, for there is no way that any other set of conditions would ever duplicate those that took place during World War II, and the danger that we might repeat Auschwitz and Hiroshima is minimal. Paradoxically, then, there would be in fact very little to be learned from these cases in political terms.

But one would be appalled to hear that there is no reason to learn about Auschwitz and Hiroshima: We have insisted over the past fifty years that the only way to maintain our humanity is by showing our barbarity. We have used these events as landmarks against which every genocide and the responses to it are measured (see, e.g., Fein 1979). More specifically, I stated above that technoscientific policies should be governed by the potential for abuses illustrated in the cases of Auschwitz and Hiroshima. So if there is something to learn about these events, what is it? To follow Bauman, what must be learned is not how unique these events were but how much they make sense within the broad framework of modernism. The appeal of the Enlightenment project (in its modernist guise) was its ability and compulsion to universalize, to overlook differences in the name of equality and liberty.

According to the Enlightenment project personal traits and backgrounds are better left outside the educational apparatus if we are to treat everyone alike and provide equal opportunities to everyone (see, e.g., Kant 1970). The same is said of the modern nation-state: Ethnic and religious differences are to be handled in the private sphere, so that the public sphere will be able to accommodate everyone equally and without interfering with anyone's liberty (see, e.g., Hegel 1967). The drive for the separation of state and church could guarantee freedom to all. And finally, politics and education were linked to the social and economic realms, for the capitalist system promised equal opportunity to anyone willing to compete and work hard; thus so that feudal and traditionalist obstacles could be overcome.

Proponents of modernity—however problematic this nonperiodizing term remains—delivered their ideological promises through the rational institutionalization of principles and tenets. One could argue that the political revolutions of the late eighteenth century followed logically from the scientific revolutions that preceded them. However, as Auschwitz and Hiroshima illustrate all too plainly and painfully, the characteristics that the technoscientific community could claim for itself could not be claimed for the entire society of which it was only one part. That is, though the technoscientific community claimed for itself value neutrality, objectivity, and truth and thus that it could become the model for modernity, the assumption was always that if the scientific community failed to police itself, then society at large would step in. Let us focus for a minute on the dialectical nature of the concept and execution of value neutrality as the regulative ideal (in Kant's sense) that inspires the technoscientific community.

The circumstances of those exterminating Jews in concentration camps and those developing nuclear bombs to thwart the threat of fascism in Europe (as the participants of the Manhattan Project saw themselves) differ in many ways. Yet reviewing some of the concerns expressed after Hiroshima may shed light on the predicament that was faced by two sets of technical communities. Though concerned with the history of Project Y, as it was also called, David Hawkins, the official historian of Los Alamos and an administrative assistant to Robert Oppenheimer, ends his report with the quote from Oppenheimer's speech of October 16, 1945 (on the occasion of receiving the certificate of appreciation for the laboratory from General Groves). This is an oft-quoted speech because of the strong language Oppenheimer used to describe his own feelings and his evaluation of the entire Manhattan Project, a speech in which he predicts that the names of Los Alamos and Hiroshima will be cursed: "If atomic bombs are to be added as new weapons to the arsenals of a warring world, or to the arsenals of nations preparing for war, then the time will come when mankind will curse the names of Los Alamos and Hiroshima" (Quoted in Hawkins 1983, 260–261). He ends his speech with a plea for a unified world that will heed the devastating lessons of the use of the atomic bomb in Japan. To some extent the world has ignored Oppenheimer's dark vision, and in fact there is still no consensus on what post–World War II policy ought to be adopted. Edith Truslow and Ralph Carlisle Smith explain how the three views of the role of the laboratory in peacetime were outlined right after the war:

> There were many opinions about what the laboratory should do in peacetime. One group, headed by one of the most senior Laboratory members, contended that the Laboratory should become a monument— it should be abandoned and its functions, if necessary or useful in peacetime, should be taken up elsewhere. Another philosophy suggested that the Laboratory abandon production of atomic weapons and conduct only peaceful research, or basic research whose application might be in the indefinite future. Others held that the Laboratory's basic purpose was research and development of atomic weapons, and that for the present at least, their design and production might or must continue. (Truslow and Smith 1983, 267)

These quotes should help illustrate to what extent the philosophical concerns over the universality or particularity of the horrors of World War II have affected the way practicing members of the technoscientific community think and work. For some, it is business as usual: "We come up with ideas and their applications are beyond our scope." Others have become more concerned with the eventual development of their ideas.

Personal Experiences

The views of Victor Weisskopf, as one of the leaders at Los Alamos, are extremely informative, if not necessarily accurate or representative. Referring to the atmosphere at Los Alamos, Weisskopf claims that "at the outset, however, we were all convinced that our work was vital, and we did not worry much about moral issues" (Weisskopf 1991, 126). And furthermore:

> In any case, at that point we were not certain that a bomb could be assembled, and some of us occasionally believed it would be [sic] never be possible.
>
> The psychological reactions to these difficulties were varied. Some of us, including myself, secretly wished that the difficulties would be insurmountable. We were all aware that the bomb we were trying to develop would be such a terrible means of destruction that the world might be better off without it. And if it was impossible for anyone to develop a nuclear bomb, there wouldn't be any danger of the Nazis having one.
>
> Then, imperceptibly, a change of attitude came over us. As we became more deeply involved in the day-to-day work of our collective task, any misgivings that we had at the start began to fade, and slowly

the great aim became the overriding driving force: We had to achieve what we had set out to do. (Weisskopf 1991, 127–128)

He goes on to explain that he is not sure the impetus for his participation rested solely on the Nazi threat as opposed to the desire to be part of an exciting project. He generalizes to say:

I can think of no one who refused to participate in the project because of a conviction that our great science would be demeaned by serving in the manufacture of the means of death and destruction. We were all actually aware that the whole civilized world was under attack by a force of the greatest evil. Still, the question loomed in our minds whether it was moral to use such a devastatingly destructive weapon even to defeat an enemy as undeniably evil as the Nazis. But this was an old question, one humankind had had to face many times during its history. There was no clearcut answer during the days of the Los Alamos project—and there still isn't. (Weisskopf 1991, 128)

Admitting that Niels Bohr "started our soul-searching and inspired us to begin to hold regular informal discussions about these issues" (Weisskopf 1991, 144), in the fall of 1944, he still wishes to differentiate between the work of the scientists in Germany and in the United States. It is as if, after the fact, members of the Manhattan Project wanted to reassure themselves and portray an image of their community that is ambivalent but not heartless:

There is no denying that constant discussions about the nature of the damage caused by fire and radiation sickness, and about the millions of deaths led to a growing numbing toward those terrible consequences. Sometimes, perhaps in the middle of the night, some of us became suddenly aware of the horror that could be caused by our work, but we were also convinced that what we were involved in was important for saving the world from the forces of Nazism, and that kept us going. I have often wondered what our attitudes would have been had we known that there was no seriously competitive Nazi effort toward a bomb. (Weisskopf 1991, 137)

Weisskopf maintains his sense of proportion and fairness, and says that "only two people, Volney C. Wilson and Joe Rotblat, left the project because of the end of the Nazi threat" (1991, 147). He adds that

Joe left also because he wanted to go to Poland to try to find his family. Tragically, he learned that his wife and her parents had been killed by the Nazis, but he did find some of his own family, who had survived the Holocaust . . . in retrospect, I have often been disappointed that, at the time, the thought of quitting did not even cross my mind. (Weisskopf 1991, 147)

The voices of some of the participants in the Manhattan Project tell about ambivalence, ambiguity, anxiety, and anguish in the face of one of the greatest feats of humanity—a feat that showed the ingenuity of humans and their appetite for merciless destruction. Otto Hahn displays a compassion and sadness that is at once genuine and pathetic, for he was in some sense indirectly the real "father" of the atomic bomb, he still worked for the German regime during the war, and he finally was captured by the Allied forces and detained in Farm Hall in the United Kingdom. Yet his voice differs little in spirit from that of Oppenheimer or many other "American" technoscientists (remember that many were German or Austrian refugees).

General Groves, the military commander of the Manhattan Project, reports in his autobiography that after hearing in Farm Hall of the news of the dropping of the bomb on August 6, 1945, Otto Hahn

> was completely shattered by this news and said that he felt personally responsible for the deaths of hundreds of thousands of people, since it was his original discovery that had made the bomb possible. He went on to tell Rittner that he had contemplated suicide when he first saw the full potentialities of his discovery, and now that these had been realized, he felt that he personally was to blame. (Groves 1962, 333)

Groves continues to quote some of the dialogue between Hahn, Heisenberg, Karl Wirtz, and Carl Friedrich von Weizsacker trying to explain away the superior American technological feat by claiming that the German physicists really did not want to succeed in building such a bomb; Hahn dissented from this view but felt grateful that the Germans did not succeed (Groves 1962, 333ff).

How does one account for the similarity between the concerns of technoscientific leaders amid a gulf of cultural, ideological, and personal differences? Is there indeed a scientific ethos, as Merton describes it, or a modernist ideal according to which careers are built and communities are organized? Or have we moved to a postnuclear era in which postmodern sensibilities can help explain differences and similarities in attitudes and orientations?

Postmodern Imagination and Responsibility

Is one left to lament the sad fact that ethics is not taught to technoscientists, that questions of personal responsibility are not foregrounded during the initiation process all technoscientists must undergo? But what morality should be taught? Would it not be just as problematic if we taught one set of values as opposed to another, and thereby tainted a technoscientific community with the ideological convictions of its culture?

Perhaps a postmodern posture or orientation can help ensure that the modernist ideal of value neutrality is not simply discarded, but that a different ideal is put in place. *However odd it may sound, I claim that postmodernism enhances the sense of responsibility of techno-scientists, just as I argue that the lack of scientific certainty increases (rather than decreases) a sense of personal responsibility.* Just because we know less or have more doubts does not mean that therefore we must all become relativists and nihilists. Instead, we can argue the reverse: In the face of general doubt, personal conviction and commitment to making choices among alternatives, perhaps differently from the existentialist mode, are the only appropriate responses (see Sassower and Grodin 1987).

If existentialism as a movement or a way of thinking has taught us something significant in the wake of World War II, it is exactly this sense of personal responsibility that accompanies choice making under particular conditions of ambiguity, no matter how absurd the conditions or how futile the choice (de Beauvoir 1991). Though a rational calculus may not readily lend itself to some situations, this need not lead one to conclude that an irrational choice has been made. What is judged to be irrational may be only arational, because the conditions under which a choice was made were themselves judged as irrational.

Postmodernism, even more than modernism, can enhance a sense of personal responsibility because of its claim that all discourses should be treated equally, that is, that no discourse—even that of techno-science—deserves to be above critical evaluation or to remain beyond methodological as well as moral reproach. This situation may be deemed problematic because according to postmodern lore, all discourses, rather than being equally open to criticism, are actually equally immune from criticism for they are at heart self-legitimating (see, e.g., Lyotard 1984, 47).

Self-legitimation as a strategy to avoid relying on a set foundation or to avoid reductionism may be perceived as an attempt to evade criticism. For example, the British Empire was powerless to impose its principles of sovereignty on the colonies when they formulated and delivered the Declaration of Independence in 1776. But self-legitimation cannot retain its power and authority and claim to negate other powers and authorities without accepting full responsibility for itself. In other words, the establishment of one's discursive legitimacy is accompanied by responsibility. Those who align themselves with a particular discourse take on the legitimation mode of that discourse and inevitably are presumed to be responsible at least for their choice of alignment if not for the internal details of the discourse.

The postmodern orientation I have in mind may differ from the one commonly portrayed by others (e.g., Félix Guattari's characterization of its political "dead-end," 1986). I know of critiques that lampoon this orientation, that claim that when the author has been erased from the text, there is no one to bear any responsibility (e.g., Rosenau 1992, 33). But the author never disappears; what disappears is only the centrality of the author's role, that is, its exclusive authority. With the disappearance of the exclusive authority of the author who dictates the meaning of the text based on intentions (God for religious texts, and humans for adherents of hermeneutics), the reader receives a larger role and thus becomes coproducer of the (interpretation of the) text. Once the participation of the reader is recognized, there is a chance that a self-realization by the reader (a reflexive moment) will accompany this process and that even a sense of responsibility will be heightened.

At this juncture we may return to some of the comments made in previous chapters about the notion of human control over the environment. Control could be understood within the Stoic formulation of dividing the world's events into those within and those outside one's control, and it could be understood within the technoscientific formulation of the age of modernity. In either case, the option of chaos, as we have seen, is shunned because it paralyzes culture. If postmodernism is to survive its faddish acceptance in Western thought, it can do so by emphasizing that its tenets come as a response to the angst of World War II, not as a means to avoid its horrors. To be an effective response, postmodernism has to be a convincing means of fostering additional personal responsibility in individuals rather than being a means of shirking responsibility. It can succeed if read in particular ways, yet it fails because most of its proponents shy away from prescribing ethical norms.

It is self-contradictory to establish criteria of behavior while decrying the follies of foundationalist philosophical traditions and models of behavior. How can we avoid the pitfalls of religious dogmas and philosophical doctrines of absolutism and yet offer principled guide-lines for human conduct? This question will be more fully answered in Chapter 5, wherein I attempt to connect the ethical ideas of Kant, Lyotard, and Emmanuel Levinas.

The postmodern broadening of otherwise narrowly defined cate-gories in their modern context may explain why critical analyses and interpretations are not focused on books alone but are equally applied to Auschwitz and Hiroshima. What the categories all require is an ac-tive imagination that has culturally been delegated to the arts (see Sas-sower 1995, Ch. 7). This use of the imagination may help realign power relations in ways yet unimagined, for a different power struc-ture must be imagined in order to be established. If the reverse hap-pens (as we shall see below), then subversion and revolt become infi-nitely more difficult, if not impossible. In short, cultural imagination is too precious to be confined to artistic productions in the age of capitalist manipulation.

My focus thus far has been primarily on the Holocaust in Europe, about which Bauman reminds us "that also *the unimaginable ought to be imagined*" (Bauman 1991a, 85). I wish now to illustrate that the same issue of the unimaginable haunts our text of Hiroshima. In the course of his correspondence with Claude Eatherly, one of the pilots who dropped the atomic bomb on Hiroshima, Günther Anders, a Vi-ennese philosopher, explains that after World War II humanity has to reconceive its convictions and behavior patterns, "for in the course of the technical age the classical relation between imagination and action has reversed itself" (Eatherly and Anders 1989, 227).

The picture I am drawing here differs from that of C. P. Snow's *The Two Cultures and a Second Look* (1964), for I do not separate two groups of intellectual workers according to disciplinary lines or the boundaries of discursive practices. The discourse of the literary critic is just as incomprehensible as that of the astrophysicist; so the simple criterion of difficulty or comprehensibility no longer works as a demar-cation between the humanities and the sciences. Perhaps the dividing feature is neither center versus periphery, public comprehension versus incomprehension, nor matter versus spirit, but the capacity and power to kill wholesale, which brings us back, then, to Auschwitz and Hiro-shima as the yardsticks with which to assess our cultural position.

After the shattering of human dignity and value in World War II, it is difficult, if not impossible, to hold onto the notion of sacredness:

Nothing is sacred anymore. This realization is part of the postmodern orientation. Modernity fought against despotic injustice and religious dogma, even though modernity was itself at times unjust and dogmatic. Postmodernity seems to fight against the injustice and terror of rationality and the dogmas of science, while being at times both rational and dogmatic. To speak, therefore, about postmodern technoscience is to speak dialectically. The dual critique of technoscience is magnified by and reinscribed with postmodern sensibilities. This speech may reflect some forms of cultural collision whose manifestation has cost lives and whose consequences launched sovereign states into conflicts and wars whose legitimacy has been assumed rather than critically evaluated (Sassower 1995).

Postmodern technoscience exemplifies Günther Anders's concern with the relationship between imagination and action. The traditional conceptual model of the transcendence of the imagination over mere human action is outdated. The model of the technical age, as Anders calls it, requires that the imagination play catch-up to the actions of humans. Human imagination is lagging behind human action by the end of the twentieth century, and the call for the imagination to transcend action is a warning call that we need the imagination to save lives, to survive into the next century, or to question power relations as they appear around us.

When Anders argues on behalf of Eatherly (who was institutionalized by the Air Force) that one should not be expected to behave normally under abnormal conditions, there is still the question: Why can't we all "see" this obvious fact of life? Do we not agree that under extreme conditions courageous action is needed, that in the face of unusual circumstances we would condemn all those who rationalized these circumstances away in the name of routine and normalization? The specter of "collective guilt" as it is applied to the German *Volk* is linked to inaction, a "normal" response of carelessness despite the dark and smelly clouds above the crematoria.

The problem of "seeing" is the problem of the imagination, though it is cognitive and social and not merely visual. In the cases of both Auschwitz and Hiroshima, the unimaginable actually occurred, that is, what we had not imagined took place: The modern and rational machinery of technoscience did not await our nightmares. So the first problem is not having the imaginative foresight to prevent the preventable. This problem is compounded once the horrors are actualized. During the events we are too overwhelmed to act, so we can only react, slowly, post hoc, retrospectively (as mentioned above in terms of power structures). The second problem then arises: How do we imagine alternative responses or reactions? Must our reactions follow the dictates of rationality, or may they defy the rules of the language game played by and adhered to so religiously ever since the Enlightenment?

Anders claims that to remain "normal," that is, to maintain the rationalizing status quo of intellectual discourse and political institutions, is a moral sham. This parallels Bauman's concern regarding the exclusion of morality from the scientific discourse: "*Science is indeed a language game with a rule forbidding the use of teleological vocabulary*" (Bauman 1991a, 170). So we must refuse the standard application of discursive rules that radically separates language games and uses these rules differently (to use Ludwig Wittgenstein's sense of the use of linguistic rules, 1958), or we must invent new rules and new games (to use Lyotard's sense of language games, 1985, 61). Either way, we could confirm the fact that a different meaning is indeed operable in the now revised language game of modernity, a language game whose vocabulary seems to allow concentration camps and atomic bombs. The imagination must come into play at this juncture and assist us in attempting to provide a different use of the techno-scientific apparatus.

The role of the imagination may be deemed inappropriate within the context of language games, unless one appreciates the current focus on the linguistic construction of reality—in the critical sense of the term. What makes the imagination crucial for the linguistic construction of reality is its ability to mix metaphors and images in ways that open different opportunities for surveying, auditing, revising, and reconstructing reality. Even when language games are not bound by disciplinary classifications, they are still bound by their own traditions and legitimation strategies. One way of overcoming traditional boundaries and the boundaries of justification is through the mixture of existing language games and their rules, and this process requires an active imagination.

When Wittgenstein quips so aptly that his "aim in philosophy" is "to shew the fly the way out of the fly-bottle" (1958, para. 309), he may not be concerned with the release of the fly from the bottle but with the very possibility that the fly has a way out. Imagining the way out, seeing the door through which one may leave one's confines, is all that is expected. Even this move of the imagination may not achieve the intended goal of liberation, as Sartre demonstrates in his 1945 play *No Exit* (1976). Perhaps an image is the most one may hope for in the name of empowerment and freedom. The very fact that the fly knows the way out is enough, for it is not the role of philosophers to get the fly out of the bottle; to be anthropocentric: The fly is free to choose to stay or leave.

Perhaps the imagination is a method, technique, or strategy with which to ensure that even transparent surfaces like those of a glass bottle do not deceive us into believing that we are not confined. Socrates' parable of the cave and its shadows must have inspired Wittgenstein, for their images are similar: Someone is on the inside not knowing that there is an outside, that there is a world different from the one in which one lives. The reader, then, becomes philosophical enough to know a way out. Just as we failed to imagine a World War II after the pain of World War I, we failed to imagine gas chambers and nuclear bombs. Must we not imagine a nuclear-free world, or a world free of hatred, prejudice, violence, and genocide?

One may laugh cynically at these pleas to imagine these impossibilities. But does one dare laugh at all after having observed that the unimaginable was imagined and actualized? Perhaps the necessity and urgency to contextualize technoscience within a postmodern orientation and attitude reflects Bauman's conviction that only social and political chaos, and not order, can ensure individual moral responsibility and dissent in the face of terror. If the postmodern condition is understood as a call for the reconstruction of ideals and dreams in the wake of linguistic deconstruction, then it is a condition that promotes the use of the imagination. An active imagination can both warn us, as does dystopian writing such as George Orwell's *1984* (1961) or Fritz Lang's film *Metropolis* (1927), against potential horrors and pave the way for a better future, as so many utopian works attempt to do.

It is true that, unlike Marxism with its Communist ideals or feminism with its vision of a successor science, postmodernism falls short, not having a blueprint for power relations, divine or human. It seems to settle for less—an orientation, an attitude, a way of life—but in fact, this is not settling for less but, rather, opening to much more because many more options could be imagined. Yet the skeptic will ask: Will new options be imagined? And by whom? And for what purpose?

Instead of deflecting the skeptic's queries, we should embrace them and set them at the threshold of our discursive exchange, so that our imagination will be guided by principles worth defending, having been made explicit and open to public scrutiny. But would this not create political instability and chaos? Is this not exactly what has been the Achilles' heel of postmodern politics, namely, its inability to formulate a sense and reality of political agency? In the face of historical analyses that argue that the rise of fascism in Europe was enhanced (if not directly caused) by political turmoil, Bauman sounds a postmodern refrain: *"The voice of individual moral conscience is best heard in the tumult of political and social discord"* (Bauman 1991a, 166).

Perhaps only when political chaos reigns is there a chance that voices different from those of the political leadership will be noticed. Some may argue that under chaotic conditions conservative trends appeal to large segments of the population (e.g., as the religious right has in the United States since the 1960s). Though this argument may be empirically true, it is also true that the only way to combat reactionary trends is through acts of "individual moral conscience," as Bauman says, or by invoking the images of the consequences of their rise to power. As Chapter 5 illustrates, there is room for individual responsibility in the face of war threats and the pressure of military and political leaderships.

5

Responsible Technoscience: A Reconstruction

The requirement to address the great calamities of our century, as it was raised in Chapter 1, is reflected in the endemic condition of ambiguity, anxiety, and anguish described in Chapters 2 and 3. The technoscientific community is in need of a postmodernist dose of flexibility and openness, as well as their consequence, responsibility, in order to ensure its political integrity—its ability to remain critical, rational, and context bound—as Chapter 4 tries to explain.

In this chapter, I begin by juxtaposing two views concerning the responsibility of scientists, the views of Oppenheimer and Teller, in order to illustrate the position into which individual scientists have been put as members of the technoscientific community before, during, and after World War II. I follow this juxtaposition with an account of the organized efforts by some groups of scientists to encourage social responsibility. I conclude the chapter with additional personal accounts, confessions, and reflections of members of the Manhattan Project to provide a set of suggestions that may be philosophical in intent but that are practical as well.

Debating Responsibility

Oppenheimer's views about his role in the Manhattan Project became a public spectacle when his security clearance was challenged at the end of World War II. Having confessed his reservations about the use of the second nuclear bomb against civilians and having become a respectable spokesperson not only for the Manhattan Project but also for the technoscientific community in general, Oppenheimer was subjected to an unprecedented level of personal scrutiny and interrogation. Portions of the transcripts of his security hearings found their way to the public record through a play by Heinar Kipphardt (who claims to have been faithful to the transcripts). Kipphardt quotes Ward V. Evans, a member of the Personnel Security Board, after being concerned with the military use of science, as saying:

> I don't know, perhaps my liberal views are outmoded; perhaps science, too, must bow to the absolute claims of the state. Now that science has become so important. At any rate, I can see two kinds of developments. The one is our increasing control over nature, our planet, other planets. The other is the state's increasing control over us, demanding our conformity. We develop instruments in order to pry into unknown solar systems, and the instruments will soon be used in electronic computers which reduce our friendships, our conversations, and thoughts to scientific data. (Quoted in Kipphardt 1968, 25)

The desire for and promise of control, as described in Chapter 2, is seen here as a double-edged sword, one that may turn on its inventors and its users. And when the control over nature is turned into a control over people as well, this may interfere with the ethos of science, as described by Merton, or the collaborative nature of the work of the technoscientific community. In Oppenheimer's words:

> People with first-class ideas don't pursue a course quite as straight as security officers fondly imagine. You cannot produce an atomic bomb with irreproachable, that is, conformist ideas. Yes-men are convenient, but ineffectual. (Quoted in Kipphardt 1968, 40–41)

The implication here is not only that nonconformism is a necessary ingredient in bringing about breakthroughs in technoscience but also that the price of individuality, dissent and confrontational debates, is essential for these breakthroughs. Edward Teller, brought in as a witness at the hearings of Oppenheimer, agrees with Oppenheimer about the need to support the work of technoscience, yet he contends that it may not be up to individual members of the technoscientific community to undermine or derail the workings of the projects in which they are involved:

> All the great discoveries had, at first, a devastating effect on the state of the world and on its image in our minds. They shattered it and introduced new conditions. They forced the world to move forward . . . If we

persevere with our work, regardless of the consequences, we shall force man to adjust to these new energies, and to put an end to that state of the world in which he was half free, half enslaved! (Quoted in Kipphardt 1968, 95)

At this juncture, Oppenheimer returns to the theme of the individual who is part of a community, and he recalls the great achievements of science and technology. The stakes have become higher as the consequences of research have had global effects. In his closing statement, Oppenheimer has this to say:

> When I think what might have become of the ideas of Copernicus or Newton under present-day conditions, I begin to wonder whether we were not perhaps traitors to the spirit of science when we handed over the results of our research to the military, without considering the consequences. Now we find ourselves living in a world in which people regard the discoveries of scientists with dread and horror, and go in mortal fear of new discoveries . . . At these crossroads for mankind we, the physicists, find that we have never before been of such consequence, and that we have never before been so completely helpless. (Quoted in Kipphardt 1968, 126)

How should one deal with Oppenheimer's view of the helplessness of technoscientists? Should one refuse to participate in projects whose consequences may turn out to be devastating to humanity? Or should one refuse, instead, to take on the responsibility of unforeseen consequences because of the involvement of military and political leaders?

According to Teller, the responsibility of scientists has a definite limit, the limit of explaining the research that has been accomplished:

> When the scientist has learned what he can learn and when he has built what he is able to build his work is not yet done. He must also explain in clear, simple, and understandable terms what he has found and what he has constructed. And there his responsibility ends. The decision on how to use the results of science is not his. The right and the duty to make decisions belongs to the people. (Teller 1964, 22)

According to Teller, scientists' responsibility is limited to the research and development of their ideas and discoveries. The use made of these ideas and discoveries is none of their concern. In this sense, Teller espouses the limits of responsibility. But what does this limit mean? He goes on to explain:

> I believe that the scientist's responsibility is limited. By being limited it is actually more difficult. What he has to do no one else can do in his stead.

And the last of his jobs, to explain clearly and objectively his results, may well turn out to be superhuman. Who can be objective? Who can separate undeniable facts from implied conclusions? Let us not expect too much from scientific objectivity. Let us be content if the scientist attempts to be honest. Let us not assume that he is unprejudiced. But let us require that he name his prejudices. Free debate between prejudiced advocates is a tortuous road toward truth. But it has proved more reliable than any straight doctrine. (Teller 1964, 22)

The debate between those claiming the full responsibility of all members of the technoscientific community and those claiming the limits of responsibility has been raging in different quarters in this century. It has been heightened because of the horrors of World War II, but its historical background may suggest how specific political contexts can shed more light on who is or should be responsible for the products of technoscience. So before we move to discuss the Atomic Scientists Movement or the Society for Social Responsibility in Science, let me review Bertolt Brecht's *Galileo*. It is an interesting coincidence worth noting, as Eric Bentley does, that

in East Berlin, 1965, Heinar Kipphardt's play about J. Robert Oppenheimer began in the setting left on stage from Brecht's *Galileo*, and the newer play "shows what society exacts from its individuals, what it needs from them," as Brecht said his *Galileo* did. However, one is struck by the extreme difference between the two main dramatic situations— that in which it is Reaction that suppresses discovery, and that in which it was inhumane to push *for* science and the *making* of a discovery. (Bentley 1966, 23–24)

Brecht is consistent in attributing to Galileo the anguish of a scientist whose labors have been misused and whose personal sense of responsibility has increased after years of lonely isolation. Brecht's Galileo says:

The practice of science would seem to call for valor. She trades in knowledge, which is the product of doubt. And this new art of doubt has enchanted the public . . . But now they have learned to doubt . . . I take it that the intent of science is to ease human existence . . . and I surrendered my knowledge to the powers that be, to use it, no, not *use* it, *abuse* it, as it suits their ends. I have betrayed my profession. Any man who does what I have done must not be tolerated in the ranks of science. (Brecht 1966, 123–124)

For Galileo, then, any time the scientific community or any of its members bows to the church or the state, tries to appease the authorities, or complies with their demands, it is a breach of the ethos of science (as Merton calls it) or the guiding principles that ought to be universally accepted and observed by all members of this community. In this context, Galileo's assistant, Andrea, visits Galileo after some years when he is under so-called house arrest, and says: "Unhappy is the land that breeds no heroes"; to which Galileo responds: "Unhappy is the land that needs a hero" (Brecht 1966, 115). So the heroic acts of technoscientists, "blowing the whistle," as we call it nowadays, dissenting from established views, suffering the wrath of people in authority (under dictatorial or democratic regimes), are sad acts. The sadness is in the fact that these acts are exemplary.

According to Bentley, there are two versions of Brecht's play, the 1938 version that addresses the work of artists under the Nazi regime ("Take care when you travel through Germany with the truth under your coat!" Bentley 1966, 15) and the 1947 one that addresses the atomic age ("The atomic age made its debut at Hiroshima in the middle of our work. Overnight the biography of the founder of the new system of physics [Galileo] read differently" Bentley 1966, 16). Bentley observes that the difference between the two versions is "*Galileo I* is a 'liberal' defense of freedom against tyranny, while *Galileo II* is a Marxist defense of a social conception of science against the 'liberal' view that truth is an end in itself" (Bentley 1966, 18–19). What is fascinating about this judgment is Bentley's concern with the ambiguity shown in the first version, where Galileo's cunning was juxtaposed against his cowardice; in the later version, Galileo is condemned by Brecht more vigorously and his entire environment is shown to be "shipwrecked." Bentley quotes Brecht to have said: "Galileo's crime can be regarded as the original sin of modern physical science . . . The atom bomb, both as a technical and as a social phenomenon, is the classical end product of his contribution to science and his failure to society" (Bentley 1966, 23). So where do technoscientists in the modern and postmodern eras stand in regard to their culpability? Must they adopt a strong view concerning their conscience qua technoscientists and qua citizens and qua human beings? Is Galileo indeed responsible for the use of nuclear bombs in Japan? Is he as responsible as Hahn or as Oppenheimer? Serres quotes Jacques Monod as saying the very day before his death:

> I used to laugh at physicists' problems of conscience, because I was a biologist at the Pasteur Institute. By creating and proposing cures, I always worked with a clear conscience, while the physicists made contri-

butions to arms, to violence and war. Now I see clearly that the population explosion of the third world could not have happened without our
intervention. So, I ask myself as many questions as physicists ask themselves about the atomic bomb. The population bomb will perhaps prove
more dangerous. (Quoted in Serres 1995, 17)

So it is not only physicists and nuclear scientists who must worry about
the consequences of their research. Instead, every technoscientist, no
matter in what area of research, might find one day that what seemed
originally to be a good idea or a worthwhile research protocol may
turn out to have devastating consequences or secondary effects that
could not have been predicted prior to the implementation of a particular program. So, one may ask, is Galileo indeed guilty for the dropping of bombs on Hiroshima and Nagasaki?

Dwight Macdonald, in his essay on the responsibility of peoples,
reminds us that when it comes to the notion of guilt, it is difficult to
draw lines of demarcation:

If "they," the German people, are responsible for the atrocious
policies and actions of "their" (in the possessive and possessing sense,
again) government, then "we," the peoples of Russia, England and
America, must also take a big load of responsibility.

We forced defeated Germany, after World War I, into a blind alley
from which the only escape was another blind alley, Nazism; this we
did by throwing our weight against socialist revolution. After Hitler
took power, more or less with our blessing as a lesser evil to revolution,
we allowed him to rearm Germany in the hopes that we could turn
him against Russia, and we used "non-intervention" to aid him and
Mussolini to overthrow the Spanish Republic in the "dress rehearsal"
for World War II. (Macdonald 1953, 11)

In these passages and many others, Macdonald reminds us to what
extent guilt must be shared more broadly than we would like to believe: It is not the burden of military leaders alone, nor of the leaders
of the technoscientific community; in addition, there are political leaders representing the concerns of the people, and thus guilt spreads
across the pond of humanity as if it had been hit in its very center by
an enormous rock. More specifically, Macdonald reminds us that in
relation to Michael Walzer's concern with effects that go beyond the
need to subdue an aggressor so as to minimize casualties at war,

500,000 European civilians died under American and British bombs . . .
It is interesting to note that, just as democracies and not the totalitarian
powers developed and used the atomic bomb, so too the British and

American air forces relied mostly on strategic bombing, directed against civilians, while the Nazis and the Russians went in more for the relatively more civilized tactical bombing, directed against troops and military installations. (Macdonald 1953, 55)

Does this mean that because the people were represented by the president and because he decided to use the bomb, the scientific community is immune from any guilt or responsibility? Macdonald reminds us that in 1945 he wrote:

> Insofar as any moral responsibility is assignable for The Bomb, it rests with the scientists who developed it and those political and military leaders who employed it. Since the rest of us Americans did not even know what was being done in our name—let alone have the slightest possibility of stopping it—The Bomb becomes the most dramatic illustration to date of the fallacy of the Responsibility of Peoples . . . Fortunately for the honor of science, a number of scientists refused to take part in the development of The Bomb . . . They reacted as whole men, not as specialists or partisans. Today the tendency is to think of peoples as responsible and individuals as irresponsible. The reversal of both these conceptions is the first condition of escaping the present decline to barbarism. The more each individual thinks and behaves as a whole man (hence responsibly) rather than as a specialized part of some nation or profession (hence irresponsibility) the better hope for the future. (Macdonald 1953, 55)

Perhaps Macdonald's advice ought to be heard once more today, because he does not allow the individual to hide behind the group, behind leaders who may fail in their judgment. And when this advice is heeded, it may point to a direction the technoscientific community may not wish its members to pursue. This predicament was at the forefront of the debates that ensued after the war. In what follows, I trace some of these concerns so as to highlight the anguish that plagued some members of the technoscientific community after the horrors of Auschwitz and Hiroshima.

Scientists for Social Responsibility

Walzer, whose approach to the question of responsibility in war situations was discussed in Chapter 1, puts the blame for the misuse of the atomic bomb in the case of Hiroshima and Nagasaki on the president of the United States: "To use the atomic bomb, to kill and terrorize civilians, without even attempting such an experiment [in negotiation], was a double crime" (Walzer 1977, 268). Walzer's analysis is based on his view that the dropping of the atomic bomb only allowed for a speedy victory, not that it was necessary for any victory at all. Utilitarian calculations, he believes, were inappropriate because they mixed civilian and military casualties. The side effect—that is, killing thousands of civilians—was no mere "evil effect" in this case for two reasons: First, it was used as a means toward achieving the "good effect" of winning the war with Japan, and second, it was not minimized but maximized (in that, rather than conventional bombs, a nuclear bomb was used, and a second bomb as well). It remains an open question whether or not the war would have been won in any case under the appropriate set of circumstances (such as conditional surrender with international mediation and intervention).

Still, what about the nuclear scientists who helped develop the bomb? Were they responsible in any sense of the term for the dropping of atomic bombs over civilian centers in Hiroshima and Nagasaki? Two groups of scientists of that period responded to these questions, and some of their thoughts were recorded. On the one hand, we have the Los Alamos group, and on the other, the German scientists who were detained in Farm Hall. Weisskopf, of Los Alamos, admits that

> discussion groups like the one we participated in had been formed in Chicago and Oak Ridge, but before the end of the war no communication was permitted between us. Indeed, most of us, myself included, had no idea that in June 1945 a group under the leadership of Leo Szilard and James Franck had submitted a written proposal to the secretary of war urging the United States not to use the bomb over an inhabited location. Had I known about this proposal, I certainly would have joined the group that framed it. (Weisskopf 1991, 148)

But he did not know, nor did he initiate a similar effort within his own community. It took some years of self-examination and reflection for Weisskopf to conclude that "on some occasion I ventured to say that the first bomb might have been justifiable, but the second was a crime" (Weisskopf 1991, 156). And this sentiment about his own role in the effort of a community became more explicit as time went by and as more revelations about the extent of the project and its consequences became more public. What about further developments, like the one refused by Oppenheimer? To this Weisskopf responds:

> I did not have to struggle with the problem of where to stand on the H-bomb development. For me such questions had long been resolved. After Los Alamos, I refused to have anything to do with nuclear weapons

development. Never again would I be tempted to join in a project that would use my scientific knowledge to fashion a weapon of mass destruction. (Weisskopf 1991, 164)

These issues were also raised and discussed by the German technoscientists, who were originally at three sites in Germany, dedicated as they were to developing nuclear weapons. Some of their leaders were under house arrest in Farm Hall. Werner Heisenberg recalls the exchange among Otto Hahn, Max von Laue, Walther Gerlach, Carl Friedrich von Weizsacker, and Karl Wirtz. While Hahn seemed upset at the fact that his discovery of uranium fission was employed in the development of a nuclear bomb, others, including Heisenberg himself, felt that as individual scientists they were all replaceable and therefore did not bear a personal responsibility for the killing of thousands of Japanese civilians; technoscience's historical march could not be stopped by any member of the technoscientific community (Heisenberg 1971, 193–195; see also Winner 1977, 68–73).

Jeremy Bernstein put together an edited version of the transcripts from Farm Hall, some of which are noteworthy and informative in the context of the present discussion. He mentions that some of the German scientists thought of suicide as they recounted their experiences in Farm Hall toward the end of the war (Bernstein 1996, 96), while making it clear that the ambivalence that ran through their accounts was at times self-serving and at times outright contrary to their rhetoric and their actions at home. For example, in Report 4, designated "Top Secret," Major T. H. Rittner reports of Operation "Epsilon" (August 6–7, 1945):

> 1. Shortly before dinner on the 6th of August I informed Professor Hahn that an announcement had been made by the BBC [British Broadcasting Company] that an atomic bomb had been dropped. Hahn was completely shattered by this news and said that he felt personally responsible for the deaths of hundreds of thousands of people, as it was his original discovery which had made the bomb possible. He told me that he had originally contemplated suicide when he realized the terrible potentialities of his discovery and he felt that now these had been realized and he was to blame. With the help of considerable alcoholic stimulant he was calmed down and we went down to dinner where he announced the news to the assembled guests. (Bernstein 1996, 119–120)

Later on, Weizsacker says: "I think it is dreadful of the Americans to have done it. I think it is madness on their part." Heisenberg retorts:

"One can't say that. One could equally well say 'That's the quickest way of ending the war.'" Hahn chimes in: "That's what consoles me" (Quoted in Bernstein 1996, 123). So just as the American group was concerned with the ending of the war and the calculus of least human destruction, while upset about their contributions to the instruments of destruction, so was the German group. In the discussion of the difference between the German and the American efforts during the war, it is clear from the Farm Hall transcripts that the German scientists, who felt they were superior scientifically to their American counterparts, needed to find an explanation for their so-called scientific defeat. Heisenberg, for one, tries to explain the situation in broad and general terms:

> The point is that the whole structure of the relationship between the scientist and the state in Germany was such that although we were not 100% anxious to do it, on the other hand we were so little trusted by the state that even if we had wanted to do it it would not have been easy to get it through. [Bernstein comments that it is questionable whether after 1942 the scientists were not trusted, especially Heisenberg himself.] . . .
>   WEIZSACKER: I don't think we ought to make excuses now because we did not succeed, but we must admit that we didn't want to succeed. If we had put the same energy into it as the Americans and had wanted it as they did, it is quite certain that we would not have succeeded as they would have smashed up the factories. [Bernstein comments: "This may be true. Witness what happened to the Auer factory. But this is a far cry from saying that we did not succeed because 'we did not want to succeed.'"] (Quoted in Bernstein 1996, 131–132)

A revisionist account of the work of German scientists under Hitler was being concocted in Farm Hall, and Weizsacker seemed to lead the group in its formulation. He went on later to say:

> History will record that the Americans and the English made a bomb, and that at the same time the Germans, under the Hitler regime, produced a workable engine. In other words, the peaceful development of the uranium engine was made in Germany under the Hitler regime, whereas the Americans and the English developed this ghastly weapon of war. (Quoted in Bernstein 1996, 154)

Report 5 (August 8–22, 1945) includes an interesting summary of the views of some of the German scientists regarding the political control of their work in the future, now that the atomic bomb had been used. Heisenberg claimed that at present all scientists were too dependent on their governments, and if they intended to have any power to prevent the use of their work for destructive purposes, they ought to seek more political influence. How exactly this would come to fruition remained an open question. (Bernstein 1996, 213)

Perhaps the ambiguity concerning the notion of guilt, understood personally and collectively (in Hannah Arendt's sense, 1963), understood against a religious or moral backdrop (in Emile Fackenheim's sense, 1970), understood historically or in terms of a utilitarian calculus (in Helen Fein's sense, 1979), especially within the conditions of warfare (where abnormal conditions may require abnormal behavior as Eatherly and Anders claim, 1989), brought about the formation of the Atomic Scientists Movement and the Society for Social Responsibility in Science. (Some would argue that the conditions that bring about wars are normal, so that the state of war itself is somehow normal, perhaps in a Hobbesian view of human nature and the state of nature.)

In describing the Atomic Scientists Movement, Donald Strickland explains that the movement "embraced two partially inconsistent roles": "The legitimate, traditional business of a scientist is the discovery and comprehension of a certain order of truth, whereas the scientist as citizen-politician is called to action, persuasion, and manipulation" (Strickland 1968, 8). This leads him to conclude as well that in the final analysis the scientists themselves have had little to do with the actual decision to drop the bombs on Hiroshima and Nagasaki. Yet in discussing the particular responsibility of scientists, Strickland quotes James Franck of Chicago, who circulated a memorandum to President Truman against the use of the bomb in the war with Japan, concerning the question of secrecy and its restriction of political as well as scientific action:

> None of the scientists objects to these regulations as long as they only bring about personal inconveniences and restrictions in mutual information which would be useful for the work. These regulations become intolerable if a conflict is brought about between our conscience as citizens and human beings and our loyalty to the oath of secrecy. That is the situation in which we scientists now find ourselves. (Quoted in Strickland 1968, 28)

But the concerns of scientists, like Franck, are not limited to their role within the scientific community and as advisers to political leaders. Instead, their concerns are embedded, according to Strickland, in the concept of "political distance" that has been adopted figuratively by the movement of scientists during this period. They tried to "keep to the periphery of politics, denying effective presence, and insisting they are specialists subject to special rules" (Strickland 1968, 78). This, of course, would be less problematic if it were not for the fuzzy distinction between the "purists" among the scientists, who claim that "good science" differs from political activism, and the "expansionists," who concede that their own community is itself a social institution. Even the latter group suggests that scientists ought to speak politically as citizens and not as scientists, thus setting up "a muddle" (Strickland 1968, 78).

Strickland outlines the main political questions that ought to have concerned scientists: First, should the atomic bomb be used against Japanese civilians? Second, could the bomb be controlled after the war, and by whom? Third, would security concerns overshadow intellectual freedom and put laboratories under the direct control of the military? Fourth, does the public fully comprehend the atomic age? And fifth, should scientists be under supervision of military and political leaders? (Strickland 1968, 101). The Atomic Scientists of Chicago put the matter in the following way:

> The scientists do not aspire to political leadership but having helped man to make the first step into this new world, they have the responsibility of warning and advising him until he has become as aware of its perils as its wonders. They have lived with the secret of the atomic bomb for several years; they thought about its future and its implications for mankind long before the rest of the world became aware of the problem. (Quoted in Strickland 1968, 101–102)

Strickland summarizes quite well the urgency and practical agenda on the minds of the members of the Atomic Scientists Movement: "the need to tell; the need to warn; the need to denounce military control and secrecy . . . it was the education of the general public in addition to the education of the political leadership of the country" (Strickland 1968, 128). In order to accomplish these tasks, and having become involved in the war effort for patriotic reasons, the group of physicists and chemists who worked in the Manhattan Project displayed a strong sense of fraternity. Moreover, the pattern of migration that brought them together in the United States did not undermine their attachments and relations with scientists who remained in Europe, even those who remained, willingly or unwillingly, under German rule. In these respects, then, this group of scientists maintained the scientific ethos as described by Merton: They valued open scrutiny of their work, fraternity, universalism (or internationalism), and freedom of thought from the state (more specifically, from the military) just as their predecessors demanded freedom of thought from the church.

The Society for Social Responsibility in Science (SSRS) declared:

> In September, 1949, a group of scientists and engineers, deeply concerned at the increasing use of science for destructive ends, met at Haverford, Pa. Believing that science and technology should contribute fully to the benefit of mankind, and never to its harm or destruction, the group organized the Society for Social Responsibility in Science . . . According to the constitution the SSRS is organized "to foster throughout the world a functioning co-operative tradition of personal moral responsibility for the consequences for humanity of professional activity, with emphasis on constructive alternatives to militarism." (SSRS 1971, 4:211)

The SSRS published a *Newsletter* in which the views and comments of leading scientists and engineers were expressed. In 1971, the society reiterated its original goals and provided some revisions. The original goals were contextualized in terms of the survival of civilization, and every scientist and engineer was asked

> (1) to foresee, insofar as possible, the results of his professional work; (2) to recognize his personal moral responsibility for the consequences of this work, irrespective of outside pressures; (3) to seek work which seems to him of benefit to mankind and abstain from that which he judges to be injurious to it; and (4) to use his scientific and technological knowledge, guided by ethical judgment, to aid government and laymen in the intelligent and humane use of the tools which science and technology provide. (SSRS 1971, 1)

Recognizing the "lonely position" of the scientist and engineer, the society tried to revise its goals and instructions, without compromising its commitment to domestic prosperity and international peace. Victor Paschkis, the chairman of the society's Committee on Review and Statement of Goals at the time, called on every applied scientist and engineer to accept the following oath:

> I vow to strive to apply my professional skills only to projects which, after conscientious examination, I believe to contribute to the goal of co-existence of all human beings in peace, human dignity and self-fulfilment [sic]. I believe that this goal requires the provision of an adequate supply of the necessities of life (good food, air, water, clothing and housing, access to natural and man made beauty), education and opportunities to

enable each person to work out for himself his life objectives and to develop creativeness and skill in the use of the hands as well as the head.

I vow to struggle through my work to minimize danger, noise, strain
or invasion of privacy of the individual: pollution of earth, air and
water, destruction of natural beauty, mineral resources and wild life.
(SSRS 1971, 2)

It should be noted that even with this oath, there are many ambiguities that may affect individual interpretations of any of the stated
goals of the SSRS's constitution or the proposed oath. Yet one must acknowledge that these statements, despite and perhaps because of their
inherent ambiguities, are courageous attempts to place moral responsibility on the shoulders of technoscientists and engineers. Instead of
turning to politicians and bureaucrats, the SSRS turns to its admittedly
self-selected membership. Perhaps what is left for us to wonder is why
its membership net has not been cast much wider, why its genealogy
and goals have not become part of the curriculum around the world.

A Provisional Code for Technoscientists

In what follows, I offer a procedural enhancement of the prescriptions
declared by the SSRS. It seems clear that the cultural context within
which we operate today keeps us underprepared for the creations of
the technoscientific community, just as they leave this community to
fend for itself against political and global odds unimaginable in the
laboratory. Given our potential myopia in the face of a new millennium, we should continually rewrite and propose a set of injunctions
for technoscientists and engineers. I do not ask for an oath nor do I
believe that Günther Anders's "Commandments in the Atomic Age"
(Eatherly and Anders 1989, 11–20) could ever be fully implemented.
Yet, even Anders's sense of commandments is procedural rather than
content driven, for he seems to be concerned with encouraging a critical debate about the process by which individuals ought to face the
dilemmas of their daily life and professional responsibility.

In what follows I also intend to respond to the kind of response—
that is, rationalization, even postmodern contextualization—that was
undertaken, for example, by the military leader of the Manhattan Project. I do not mean that General Groves was malicious and evil, nor
that he was a dupe of his own ideological rhetoric; instead, a deeper
concern is at issue. Generals will remain in charge of military budgets
and efforts that may lead to mass destruction. These efforts will al-

ways be carried out in the name of deterrence and lasting peace, but the generals in charge will never hesitate to use the arsenals under their control. They, more than their technoscientific counterparts, will have the "last word" and will be able to control the flow of information to sway Congress or the president. So as a springboard to my recommendations, let us review some of General Groves's views.

General Leslie R. Groves was obsessed with secrecy: "Compartmentalization of knowledge, to me, is the very heart of security. My rule was simple and not capable of misinterpretation—each man should know everything he needed to know to do his job and nothing else." Contrary to the scientists themselves, he believed that

> adherence to this rule not only provided an adequate measure of security, but it greatly improved over-all efficiency by making our people stick to their knitting. And it made quite clear to all concerned that the project existed to produce a specific end product—not to enable individuals to satisfy their curiosity and to increase their scientific knowledge. (Groves 1962, 140)

It is interesting to note that Groves believed that German scientists did not support their government in the war effort and only obtained money in order to further their own research. In order to increase their funding, they also claimed that German science was far behind American science (Groves 1962, 230). He summarizes the view, expressed by many after the war, that the German scientists "seemed to have come to feel that, while there was some hope of producing energy from a uranium pile, it was unlikely, if not entirely impossible, that a workable weapon could be developed" (Groves 1962, 244). Was the implication that success in science depended on the patriotism of the members of the technoscientific community?

As for the question of using the bomb after the Nazi defeat, Groves held that the issue was not one country's surrender but the end of the war, so the use of the bomb against Japan made perfect sense: It would "save untold numbers of American lives" (Groves 1962, 264–266). The intent was to bring "home to the Japanese leaders the utter helplessness of their position. When this fact was re-emphasized by the Nagasaki bombing, they were convinced that they must surrender at once" (Groves 1962, 319). Here the consent of the technoscientific leadership is not even mentioned. Once they had worked out the details of the weapon, it was up to General Groves to decide what to do with it.

As if to underscore this view of the different role of technoscientists and military personnel, he recalls Enrico Fermi's suggestion the night before the Trinity experiment that

> after all it wouldn't make any difference whether the bomb went off or not because it would still have been a well worth-while scientific experiment. For if it did fail to go off, we would have proved that an atomic explosion was not possible. (Groves 1962, 297)

So one way or another, the technoscientific community would be satisfied that an actual experiment was afoot. The worst that could happen would be that indeed the bomb did work, and then it would no longer be under the control of the experiment teams. Does this mean that one ought to do technoscience for technoscience's sake? I am not sure this motto remains as straightforward as it may have sounded a generation ago. And perhaps this suggests we must follow Serres's advice about the role of philosophy: "Not only must philosophy invent, but it invents the common ground for future inventions. *Its function is to invent the conditions of invention*" (Serres 1995, 86). He goes on, more specifically, to address the concerns discussed here:

> The great problems of our era, since the dawn of Hiroshima, have to do with the whole set of relationships between the law and science. We must reinvent the place of these relations; we must therefore produce a new philosophy, so that lawyers can invent a new system of laws, and perhaps scientists a new science. As a result, this critical era no longer consists of giving philosophy the right to judge everything—a regal position from which it makes rulings right and left on everything—but the responsibility to *create*, to *invent*, to produce what will foster production, to invent or express a system of laws, to understand and apply a science. (Serres 1995, 137)

As an attempt to invent, in Serres's sense, and as an attempt to suggest some safeguards for our work, one may reconsider Anders's "Commandments in the Atomic Age" as an inspiration for my own set of suggestions. In general, his commandments are prescriptive (and thus mine parallel them), but they are are more broadly construed than mine. He specifies what one must think about upon awakening every morning: He advises one to begin by thinking about the atom as opposed to a stable reality and to continue with a widening of one's sense of time. He entreats everyone not to be a coward and to make the atomic reality part of her or his "business" to avoid being fooled by experts and their alleged successful experiments. He ends with a warning against believing that atomic production can cease or that just because atomic bombs are used as a deterrent they will never be used again against an enemy (Eatherly and Anders 1989, 11–20). My own suggestions do not instruct people on what and how to think but are more concerned with how people react to specific situations in which they find themselves. Moreover, my suggestions are not limited to atomic bombs but also encompass other situations of the techno-scientific community (thereby accounting also for those working on Zyklon B for concentration camps). In supplementing and revising some of his ideas, I endorse Anders's commandments.

The following set of suggestions assumes a certain level of personal autonomy, an overwhelmingly difficult accomplishment. This means both psychological and intellectual maturity as well as a financial independence that allows one to quit a job, for example, because of fundamental disagreements about integrity and honesty. Nonetheless, I proceed in order to have a starting point for further debate. That is, I hope these suggestions are useful enough to induce politicians and citizens alike to provide the material and psychosocial conditions under which these suggestions could be easily followed by every member of society. Finally, I take it for granted that each of these suggestions deserves further elaboration that is missing here:

1. *Ask.* When signing up for a project/protocol, ask as many questions as possible. For example, ask who will participate? What is the goal? Who will benefit? What side effects are anticipated? Who is financing the project/protocol? Who is in charge, and why? Are there any guarantees for the effective implementation of the project/protocol? If these questions are not answered to your satisfaction, do not join the team.

2. *Answer.* Whatever answers you receive, accept them provisionally and skeptically; try to test and falsify them; even when the answers seem acceptable, keep on looking for alternative answers from sources outside your community.

3. *Imagine.* Try to imagine the unimaginable so as to add to your set of questions and to your alternative answers. This way you may ensure an ongoing broadening of your conceptual horizon. In the course of his correspondence with Claude Eatherly, one of the pilots who participated in the atomic bombings, Günther Anders, a Viennese philosopher, explains that after World War II humanity has to recon-

ceive its convictions and behavior patterns, "for in the course of the technical age the classical relation between imagination and action has reversed itself" (Eatherly and Anders 1989, 227). That is, the imagination does not always precede technology; rather, at times it plays catch up.

4. *Think*. Whenever the logic of a particular situation or paradigm seems compelling and you are ready to comply with its internal consistency, make sure to push the logical boundaries to the breaking point. This way implicit pitfalls may become explicit.

5. *Quit*. Be ready to quit at any given moment, not because your level of competency has been reached but because you may discover something offensive about the project/protocol. This way you may avoid becoming an accomplice to preventable atrocities. The attitude that accompanies someone ready to quit at any given moment is of sufficient detachment to ensure thorough criticism (even self-criticism).

6. *Dare to be skeptical*. Never think of projects/protocols as sacred. They are devised and executed by humans, and therefore their fallibility is inherent in their humanity. Instead of avoiding errors, embrace them as important lessons without which knowledge remains limited.

7. *Divulge*. Never think of projects/protocols as secrets. Secrecy shields projects/protocols from criticism, and without criticism they are bound to become dogmatic. Dogmatism breeds entrenched power relations, and those manifest themselves in dictatorships. Blow the whistle even if you are unsure of yourself; if you are right, many lives can be saved; if you are wrong, you may need a new job (this is a small price to pay for organized self-policing, compared to governmental surveillance). It should be noted here that it is crucial to distinguish between those whistle-blowers who are concerned with wrongdoing and those tale-tellers whose allegations are groundless.

8. *Indulge*. Whenever possible, stray beyond the confines of your specialty, so as to learn from others and to examine your own knowledge base. When doing so, you may find out not only what others are doing in the project/protocol but also what you are supposed to be doing.

9. *Explain*. Never think that what you do is beyond laypeople's comprehension. Spend as much time as it takes to articulate in detail and in plain language whatever it is that you are doing, so as to break down any potential barrier between your work and others' scrutiny.

10. *Remember*. Remember that whatever you are, you are still a member of the community, a citizen with rights and duties no different from those of others. Your temporary membership in the scientific community is a responsibility, not a privilege; it requires you to be accountable to more, not fewer, people around you.

Having listed these procedural suggestions for a provisional code for members of the technoscientific community, I still wonder: Can we hope to prevent the atrocities of World War II from ever happening again? I doubt that another set of procedural rules and suggestions can ensure the moral integrity of which Merton speaks or overcome the predicaments of war in Walzer's sense. Perhaps a broad-based educational system that instills these values in technoscientists-in-training may help us; perhaps the horrors of the Holocaust and the atomic bombs will assist them in realizing the stakes of their activities. In the face of increasing uncertainty concerning the research and development of technoscientific projects, everyone's personal responsibility ought to be enhanced, not diminished (see, e.g., Sassower and Grodin 1987). Would a community either carelessly or deliberately insist on repeating its errors?

Perhaps the way to answer this question is not limited to the provisional codes I have formulated; perhaps more fundamental rethinking about ethics must be undertaken (as I do in Chapter 6) in order to more fully appreciate the significance and limits of these provisional behavioral recommendations. But before we move on to Chapter 6, I suggest we recall some of the concerns voiced in literary texts, especially plays, that both induced and reflected the cultural context under which the evaluation of ethical commitments can be best examined.

|6

The Price of Responsibility: From Personal to Financial

This chapter attempts a difficult feat: to retain a postmodern orientation in the ethical realm despite the limits of postmodernism. In other words, I try to walk a tightrope that spans the abyss between absolutism and relativism. As Bauman so eloquently says:

> The ethical paradox of the postmodern condition is that it restores to agents the fullness of moral choice and responsibility while simultaneously depriving them of the comfort of the universal guidance that modern self-confidence once promised. Ethical tasks of individuals grow while the socially produced resources to fulfill them shrink. Moral responsibility comes together with the loneliness of moral choice. (Bauman 1992, xxii)

As we have seen in the previous chapters, the questions that loom over our culture for the next century, as well as for the next millennium, are not limited to a set of ethical principles or moral prescriptions. On the evidence of the events of World War II, we can sadly conclude that the implementation of principles and prescriptions by political leaders falls short of the intentions of individuals and parliaments. For example, despite a comprehensive set of fully articulated and recorded ethical guidelines, Nazi Germany undertook some of the most heinous experiments on human subjects in concentration and labor camps (Annas and Grodin 1992). More generally, the optimism that accompanied technoscientific research and development prior to World War II has been displaced by a heavy dose of anguish and pessimism. As Oppenheimer told his listeners at the Massachusetts Institute of Technology in 1947:

> But that I should be speaking of such general and such difficult questions at all reflects in the first instance a good deal of self-consciousness on the part of physicists. This self-consciousness is in part a result of the highly critical traditions which have grown up in physics in the last half century, which have shown in so poignant a way how much the applications of science determine our welfare and that of our fellows, and which have cast in doubt that traditional optimism, that confidence in progress, which have characterized Western culture since the Renaissance. (Oppenheimer 1955, 82)

Does the shift from optimism to self-consciousness change the cultural role and practice of members of the technoscientific community? How have their experiences transformed them individually and as a group? Oppenheimer continues:

> Despite the vision and the far-seeing wisdom of our war-time heads of state, the physicists felt a peculiarly intimate responsibility for suggesting, for supporting, and in the end, in large measure, for achieving the realization of atomic weapons. Nor can we forget that these weapons, as they were in fact used, dramatized so mercilessly the inhumanity and evil of modern war. In some sort of crude sense which no vulgarity, no humor, no overstatement can quite extinguish, the physicists have known sin; and this is a knowledge which they cannot lose. (Oppenheimer 1955, 88)

So it makes sense to shift the discussion here from ethics in general or from the anguish of those involved in the production of nuclear weapons to the responsibility of individuals. As Bauman says:

> All in all, in the postmodern context agents are constantly faced with moral issues and obliged to choose between equally well founded (or equally unfounded) ethical concepts. The choice always means the assumption of responsibility, and for this reason bears the character of a moral act. Under the postmodern condition, the agent is perforce not just an actor and decision-maker, but a *moral subject*. The performance of life-functions demands also that the agent be a morally *competent* subject. (Bauman 1992, 203)

This chapter begins with an artistic rendition of the situation in which our culture finds itself since the war, the existential predicament that permeates every facet of our life. It continues with the problem of defining guidelines, or matrices, acceptable to the culture as a whole, accounting along the way for Lyotard's view of justice. Lyotard's discussion of justice is used as a starting point in relation to which the ideas and concerns of Kant, Simone de Beauvoir, and Emmanuel Levinas can be understood. What is at stake is not a set of behavioral commandments but, rather, a warning about the limits of our legal imagination.

The chapter ends with a brief consideration of the financial costs of complying with the ethical tenets most fruitful for postmodern technoscience.

Technoscientific Angst: The Abyss of Lunacy

The Swiss playwright Friedrich Durrenmatt describes the concerns voiced so far in this book in his play *The Physicists*. The main characters all represent scientists who find themselves (voluntarily or not) in an insane asylum. Here is a crucial moment of self-reflection and personal revelation, set after World War II:

MOBIUS: There are certain risks that one may not take: the destruction of humanity is one. We know what the world has done with the weapons it already possesses; we can imagine what it would do with those that my researches make possible, and it is these considerations that have governed my conduct. I was poor. I had a wife and three children. Fame beckoned from the university; industry tempted me with money. Both courses were too dangerous. I should have had to publish the results of my researches, and the consequences would have been the overthrow of all scientific knowledge and the breakdown of the economic structure of our society. A sense of responsibility compelled me to choose another course. I threw up my academic career, said no to industry, and abandoned my family to its fate. I took on the fool's cap and bells. I let it be known that King Solomon kept appearing to me, and before long, I was clapped into a madhouse.

NEWTON: But that couldn't solve anything.

MOBIUS: Reason demanded the taking of this step. In the realm of knowledge we have reached the farthest frontiers of perception. We know a few precisely calculable laws, a few basic connections between incomprehensible phenomena and that is all. The rest is mystery closed to the rational mind. We have reached the end of our journey. But humanity has not yet got as far as that. We have battled onwards, but now no one is following in our footsteps; we have encountered a void. Our knowledge has become a frightening burden. Our researches are perilous, our discoveries are lethal. For us physicists there is nothing left but to surrender to reality. It has not kept up with us. It disintegrates on touching us. We have to take back our knowledge and I have taken it back. There is no other way out, and that goes for all of you as well.

EINSTEIN: What do you mean by that?

MOBIUS: You must stay with me here in the madhouse. (Durrenmatt 1964, 80–81)

Mobius (perhaps meant to represent Oppenheimer?) argues for the sanctuary of the madhouse; he recommends staying there rather than trying to reverse the tide of technoscientific progress and the implementation of instruments of mass destruction. Whether the knowledge in his possession is vanity or a personal disclosure, Mobius is convinced that it is revolutionary (on the scale of nuclear physics, one is led to believe). So it stands to reason that he is frightened by the prospects of what his knowledge may entail, what it may bring about, what it may accomplish. Instead of risking the consequences of his ideas, Mobius chooses to escape, to find refuge in the insane asylum. There he can be ignored; there he is safe from the temptations of both the university and industry; there he can be safely overlooked by those eager to harness his ideas to a technology that will change the face of the earth, that will undermine the status quo in the name of progress but with no moral guidance or social guarantees.

Durrenmatt considers the insane asylum to be a safe haven for revolutionary thinkers, a place to hide from their own ideas and discoveries. But is this indeed the only way to escape the existential predicament of individual thinkers, technoscientists, and artists? According to Bauman, there is a perhaps less secluded alternative:

> Aware of the danger (it is this awareness that shows up in the admission of the authority of the supra-individual standards), the contingent person *knows* that she "walks a tightrope over an abyss, and is therefore in need of a good sense of balance, good reflexes, tremendous luck, and the greatest among them: a network of friends who can hold her hand." Contingency needs friendship as an alternative to the lunatic asylum. (Bauman 1991b, 246)

For Bauman, then, the answer to anxiety and anguish is not the lunatic asylum but, rather, a community of friends who support and cajole, who care, who are critical, who are supportive of one's initiative to be a "moral subject," that is, a responsible individual. To what sort of community of friends does one belong? Is it the members of one's professional and institutional organization? Or is it a community outside the technoscientific community? And what kind of responsibility should the working member of the technoscientific community assume?

The Limits of Ethical Pluralism

Oppenheimer struggles with the notion of responsibility and tries to confine the technoscientist's responsibility to the internal workings of that community, rather than spreading it throughout the culture and thereby diffusing its focus and forcefulness:

> The scientist should assume responsibility for the fruits of his work. I would not argue against this, but it must be clear to all of us how very modest such assumption of responsibility can be, how very ineffective it has been in the past, how necessarily ineffective it will surely be in the future. In fact, it appears little more than exhortation to the man of learning to be properly uncomfortable, and, in the worst instances, is used as a sort of screen to justify the most casual, unscholarly and, in the last analysis, corrupt intrusion of scientists into other realms of which they have neither experience nor knowledge, nor the patience to obtain them. (Oppenheimer 1955, 91)

Instead of a level of responsibility that is diffuse and ineffective, instead of pretense that accomplishes little, Oppenheimer suggests that his community of investigators should pay more attention to their professional life:

> The true responsibility of a scientist, as we all know, is to the integrity and vigor of his science. And because most scientists, like all men of learning, tend in part also to be teachers, they have a responsibility for the communication of the truths they have found. This is at least a collective if not an individual responsibility. That we should see in this any insurance that the fruits of science will be used for man's benefit, or denied to man when they make for his distress or destruction, would be a tragic naivete. (Oppenheimer 1955, 91)

Oppenheimer confines the responsibility of members of the technoscientific community without abdicating it. He seems to suggest that personal responsibility is assumed when a sufficiently alert community expresses its collective responsibility. But is this indeed true? Is his community self-conscious enough to ensure the responsible training of its members? What attitudes should be fostered by members of the technoscientific community? In a speech given in 1950, Oppenheimer reminds his audience of the grounds of science:

> What are the lessons that the spirit of science teaches us for our practical affairs? Basic to them all is that there may be no barriers to freedom of inquiry. Basic to them all is the ideal of openmindedness with regard to new knowledge, new experience and new truth. Science is not based on authority. It owes its acceptance and its universality to an appeal to intelligible, communicable evidence that any interested man can evaluate. (Oppenheimer 1955, 114)

Oppenheimer goes on to explain that for his fellow researchers it is "honorable to doubt," but that this is an attitude different from a "deliberate attempt of skepticism" (Oppenheimer 1955, 114–115). The level of agreement, if not consensus, among technoscientists is achieved because they support each other's findings while examining them as closely as possible. They attempt to figure out the truths about the universe they live in rather than to undermine the knowledge base acquired by their predecessors. All of this leads Oppenheimer to observe "how limited, how inadequate, how precious is this condition of life; for in his relations with a wider society, there will be neither the sense of community nor of objective understanding" (Oppenheimer 1955, 138). But this brings again to mind the loneliness—the sense of anguish and unavoidable angst—that permeates the life of the technoscientist.

Bauman's sentiment that friendship can be an alternative to the insane asylum echoes Oppenheimer's experience and speeches. In neither case is this sentiment confined to the existential predicament of individuals or to some overall cultural angst that eludes individuals. Instead, I see their critical descriptions as warnings whose role is to stimulate a more energetic involvement by those who care about the future (because they have children or because they believe there is value in the survival of the human race). But if the technoscientific community is isolated from the society in which it operates, and if within that society many other communities and subcultures have their own agendas, how can policy makers ever achieve workable solutions? How, in the midst of debate, conflict, and disagreement, can we offer guidelines that would satisfy us all? In Bauman's words:

> With pluralism irreversible, a world-scale consensus on world-views and values unlikely, and all extant *Weltanschauungen* firmly grounded in their respective cultural traditions (more correctly: their respective autonomous institutionalization of power), communication across traditions becomes the major problem of our time . . . The problem, therefore, calls urgently for specialists in translation between cultural traditions. The problem casts such specialists in a most central place among the experts contemporary life may require. In a nutshell, the proposed specialism boils down to the art of civilized conversation. (Bauman 1987, 143)

But what is a civilized conversation, and by whom is it held? Is the emphasis on the nature of the conversation itself or on its contributors?

According to Hans-Georg Gadamer, a conversation is the process of understanding, a process that contains elements of translation as interpretation (Gadamer 1989, 383–389). Translation, as I have argued elsewhere (1995, Ch. 2), is a complex process because of the inherent indeterminacy identified by Willard Van Orman Quine. Moreover, translation brings to bear a set of presuppositions that contextualize the process and enable its completion. It seems that Gadamer does not take into consideration the need to deal with issues outside the context of the conversation itself. And perhaps there is some virtue in excluding exogenous variables as much as possible, so that the immediacy of the moment of conversation can overcome obstacles.

Yet the virtue of the moment, as we have seen in the case of the Manhattan Project, when the frenzy to accomplish a prefigured task overwhelms all other considerations, may exact a price of despair and devastation. If the people at Los Alamos did indeed commit a "sin," as Oppenheimer claims, it is not because they failed to converse with each other at all; instead, it is because their conversation was insulated from other conversations, because their discourse was not translated into the broader discourse of war and peace in Europe and Asia. Conversation as such may provide the necessary conditions but it does not provide the sufficient conditions under which broader understanding of the technoscientific discourse can be achieved. Perhaps this is because the interpretive element of the conversation of collaborators seems to need no translation, and so receives none, as if everything is clear. In what follows, I examine Lyotard's view of justice in light of de Beauvoir and Kant as a means to infuse the technoscientific discourse with elements that may turn out to be constitutive and not exogenous.

From Objectivity to Case-by-Case Judgment

The particular strategy used by Lyotard—judging case by case—is helpful when using the judicial system as a heuristic device and as an approximation for eventual formulations of rules and laws. All judgments, whether legal or technoscientific, face similar constraints and problems in their simultaneous attempt to adhere to rules and to break them, to remain connected to an "objective reality" while proposing to construct and formulate it in creative new ways. Moreover, qua judgments, they are subject to the follies described by Epictetus and the Stoics: They are human constructs that interpret nature and are within our control to reformulate.

The particular choice Lyotard makes of Kant's notion of a regulative ideal is a means by which to avoid the absolutist–relativist judgment dichotomy by taking the absolute as the distant case. If the modernist mode of thinking has been absolutist regarding truth (foundationalism) and the postmodernist relativist regarding rationality (nonfoundationalist), then a preference for the latter has to be explained in practical terms (i.e., as a research program). For example, can postmodern relativism be flexible enough without being so lax as to condone the Holocaust? This question transforms an existentialist concern into a cultural judgment, one informed by the particular attitude expressed in the work of Simone de Beauvoir.

Appreciating existentialism as "a philosophy of ambiguity," de Beauvoir scoffs at philosophers' attempts to mask the "tragic ambiguity" of the human condition. In their stead, she suggests that we should "try to assume our fundamental ambiguity." Citing our mastery of the atomic bomb, which has the power to destroy us, she says: "It is in the knowledge of the genuine conditions of our life that we must draw our strength to live and our reason for acting" (de Beauvoir 1991, 9). The genuine conditions about which she speaks are those that lead humans to anxiety and anguish, that fully recognize the ambiguity of our existence. What does she exactly mean by defining the human condition as ambiguous? How does that follow the path discussed in earlier chapters that leads from ambiguity to anxiety and then to anguish? De Beauvoir begins by recalling in general the standard worldviews with which humans surrounded themselves. Perhaps she echoes here Freud's description of the progress of civilization, keeping an eye on the traditional and modernist uses of religious beliefs and doctrines. As she says: "After having lived under the eyes of the gods, having been given the promise of divinity, one does not readily accept becoming simply a man with all his anxiety and doubt" (de Beauvoir 1991, 46–47). As Gellner warns, it almost hurts to lose one's ordered universe, one's organized and institutionalized environment. But however ordered one's universe has become, its very organization has been suspect all along. According to de Beauvoir:

> The fundamental ambiguity of the human condition will always open up to men the possibility of opposing choices; there will always be within them the desire to be that being of whom they have made themselves a lack, the flight from the anguish of freedom; the plane of hell, of struggle, will never be eliminated; freedom will never be given; it will always

have to be won; that is what Trotsky was saying when he envisaged the future as a permanent revolution. (de Beauvoir 1991, 118–119)

De Beauvoir's adherence to Trotsky's permanent revolution is a call for the continued engagement with ethics in general and with ethical dilemmas in particular. But before we examine the details of her recommendation, it should be noted that for her, unlike for other existentialist thinkers,

> the notion of ambiguity must not be confused with that of absurdity. To declare that existence is absurd is to deny that it ever be given a meaning; to say that it is ambiguous is to assert that its meaning is never fixed, that it must be constantly won. (de Beauvoir 1991, 129)

With these words, de Beauvoir seems the quintessential postmodern intellectual: distinguishing between possible responses to a situation: An individual may act responsibly, or he or she may respond nihilistically. Because evil is neither an error nor an impossibility under the condition of ambiguity, it must be accounted for, dealt with, and if possible overcome. The atomic bomb did not just happen to us; we took an active role in developing it to a stage wherein it was used against civilians; we cannot deny its evil effects; instead, we must confront these effects and those that may follow them, and hope to delineate a future in which evil effects are minimized. She continues:

> An ethics of ambiguity will be one which will refuse to deny *a priori* that separate existants can, at the same time, be bound to each other, that their individual freedoms can forge laws valid for all. (de Beauvoir 1991, 18)

In some bizarre way, and despite the history of ethical theories from Aristotle through Kant and into the present, de Beauvoir is comfortable in asserting the prominence and even privileged position that ought to be accorded to existentialist ethics:

> Therefore, not only do we assert that the existentialist doctrine permits the elaboration of an ethics, but it even appears to us as the only philosophy in which an ethics has its place. (de Beauvoir 1991, 33–34)

If de Beauvoir's existentialist ethics, or an "ethics of ambiguity," as she calls it, is tantamount to a postmodern ethics, then it has the potential of empowering not only individuals but communities as a whole, because the conditions of ambiguity may differ from one person to another, yet they permeate and are readily detected in all of our existence; no one is immune from the anguish that results from the deep ambiguity we feel about our very existence. At this juncture, Lyotard's notion of justice comes into play.

Despite his critique of Kant's presuppositions about totality, unity, and finality, that is, about the uniformity of the membership in society and about its discipline and set of beliefs (Lyotard and Thébaud 1985, 94–95), Lyotard finds some value in Kant's view of ethics. Kant's Idea, as Lyotard explains, "is a sort of horizon that performs a sort of regulatory role with respect to action . . . it is simply a pushing to the limit, the maximization of a concept. And the concept here is that of freedom, that is, of reason in its practical use" (Lyotard and Thébaud 1985, 46). So Lyotard right away combines the most abstract of ideas, the Kantian regulative ideal (or Idea) of justice with the most practical of rules of conduct, the way people should interact daily.

But if what is at issue is our actions, and if, following Lyotard, "there is no knowledge of practice," then what do we do? He begins by reminding us that when one speaks of society, "there are contingencies; the social web is made up of a multitude of encounters between interlocutors caught up in different pragmatics." And this "condition of ambiguity," to use de Beauvoir's term, leads Lyotard to proclaim: "One must judge case by case" (Lyotard and Thébaud 1985, 73–74). But if we are set up to see the horizon, the limits of our ability to judge, then how is our case-by-case focus supposed to resolve our ambiguities, anxieties, and anguish? In the case of ethics and politics, continues Lyotard, the Kantian Idea "allows us, if not to decide in every specific instance, at least to eliminate in all cases . . . decisions, or to put it in Kant's language, maxims of the will, that cannot be moral" (Lyotard and Thébaud 1985, 73–74).

Whether it is the demarcation between chaos and order in Gellner's sense or between the pleasure and reality principles in Freud's sense or between absurdity and ambiguity in de Beauvoir's sense, one thing the Kantian regulative ideal can accomplish: It can judge where one has transgressed beyond the boundaries of moral conduct. The fact that Lyotard's own prescription limits one's judgment to a contextualized situation or to the logic of the situation in feminist or Popperian terms (see Sassower 1994) does not rob it of its ethical commitment. Instead, it acknowledges the multiplicity of perspectives and individuals, their different discursive strategies, and their divergent alliances and memberships while insisting all along on eliminating evil or blatant immoral behavior:

> Yes, there is a multiplicity of justices, each one of them defined in rela-
> tion to the rules specific to each game . . . Justice here does not consist
> merely in the observance of the rules; as in all the games, it consists in
> working at the limits of what the rules permit, in order to invent new
> moves, perhaps new rules and therefore new games. (Lyotard and
> Thébaud 1985, 100)

In proposing to confine judgment and justice to individual cases, Lyotard follows de Beauvoir. He demands the kind of engagement she does; he expects that a (Kantian) regulative ideal will inspire us to debate the limits of rules, to revise them constantly, and to appreciate the temporality of all our judgments. Unlike those who may chide the existentialists and the postmodernists for their lack of commitment or foundation, both de Beauvoir and Lyotard argue that this posture is the ultimate commitment: The appeal is immediate and ever pre-sent to a reasoned action that must be legitimized, that must ensure human dignity in the face of ambiguity. De Beauvoir does not lack an appreciation of human anguish as many of her existentialist counter-parts describe it; rather, she finds a way to respond to the ambiguity that surrounds the human condition with a productive or pragmatic mind-set.

In the case of the nuclear community at Los Alamos or in the case of the technoscientists who perfected chemicals for gas chambers, the context or the situation within which judgments are to be made can-not be too narrowly defined. The goal is not to justify any and all ac-tions but to push responsibility to its limits, so as to expose the futility of blaming "others" or of expecting not to feel repercussions for one's decisions. If the individual, whether technoscientist or artist, believes that regulative ideals alone are rules of conduct, then transgression is easily justified by a wish to be a rebel or a nonconformist. But if these ideals are understood as parameter settings or as heuristic devices with which to reconfigure a situation and the variables that contribute to its very definition, then there is room for judgment. Judgment in the face of ambiguity requires reflection. Reflection, as we have seen from Oppenheimer's experience, leads to a level of self-consciousness that is conducive of assuming personal responsibility over the process and ef-fects of the actions of one's community. And here the strictures of Lev-inas are relevant.

Back to Intersubjectivity: Levinas

Lyotard attributes to Levinas the focus of contemporary philosophy on ethical questions (though Lyotard is nonetheless a critic of Levinas's thought). Levinas takes us all the way back to our concerns with the alleged facts of technoscience, that is, to the references one may make about reality as such (as if Kant's view has never been challenged). He begins by asking: "What is the relation between justice and truth?" and right away juxtaposes two seemingly incommensurable discourses (of science and of ethics). He goes on to explain that "truth is in effect not separable from intelligibility; to know is not simply to record, but always to comprehend." And when we comprehend, when we compile what we know and examine it in light of past experiences (of our own or those of others), we require "a theoretical justification or a reason." And when we require this sort of engagement, we are bound to agree that "knowing becomes knowing of a fact only if it is at the same time critical, if it puts itself into question, goes back beyond its origin" (Levinas 1969, 82).

Levinas gently carries us back to the anguish of technoscientists, to the sense of despair they may feel not only because of the ambiguity of the human condition as described by de Beauvoir but also because of the linkage between our knowledge and the limits of our modes of inquiry, our inherent incompetence (perhaps in Holton's sense, 1996). If the knowledge of reality—its facticity—is unable to anchor us as individuals, then what would ensure our place in the world? What would help us believe in ourselves and our capacity to survive?

Levinas deviates from Martin Buber's concern with an I–Thou relationship that is at once immediate and in its immediacy opens the potential for reciprocity (Levinas 1985, 98). Though he agrees with the immediacy of vision and visualization of another human being (an embodied Other), and though he insists on the supremacy of the face as a medium for engagement and an object of contemplation and full absorption, he suggests the primacy of individual responsibility without any recourse to its effects: "The relation to the face is straightaway ethical. The face is what one cannot kill" (Levinas 1985, 87). It is as if he is saying that once a human face is attached to the dropping of a nuclear bomb, one cannot escape the responsibility for murder, for one deliberately avoids the visual contact that ought to have guided the decision to drop the bomb. Moreover, Levinas claims that if one

were to take one's responsibility seriously and maintain one's commitment to that responsibility, one's personal engagement with another
would extend to a whole community, something unclear in either
Buber's work or in other existentialist writings of the period. In Levinas's words:

> ... responsibility prior to all commitment ... responsibility that oblig-
> ates us to others ... my responsibility for the deeds, the fortune and the
> misfortune that are due to the freedom belonging to others and not to
> myself. Or, more simply, by human fraternity. (Levinas 1994, 127)

Individuals with a sense of personal agency combine into a community of moral agents whose sense of responsibility determines the well-
being of the community as a whole.

In describing one of his works, Levinas asserts that responsibility is
the "essential, primary and fundamental structure of subjectivity. For
I describe subjectivity in ethical terms." To be human, for Levinas, is
to be ethical, responsible for one's actions and one's relations to other
human beings; he continues:

> Ethics, here, does not supplement a preceding existential base; the very
> node of the subjective is knotted in ethics understood as responsibility.
>   I understand responsibility as responsibility for the Other, thus as re-
> sponsibility for what is not my deed, or for what does not even matter
> to me; or which precisely does matter to me, is met by me as face.
> (Levinas 1994, 95)

In this fashion, then, Levinas extends, if not reverses, de Beauvoir's view
about existential ethics. To begin with, ethics is an initial and constitutive element, not a by-product, in existential thought. Moreover, the
provocation of ethical accountability is a form of fastening responsibility to each individual who is self-conscious of any sort of human
interaction. And finally, responsibility is not limited to one's own behavior; rather, it impels each one to feel responsible for another, thus
thrusting angst-ridden individuals outside their self-absorption toward
others who do not expect attention outside their own self-reflection.
Before long, every person is involved with another to form a provisional community of responsible individuals. As he explains: "Justice
only has meaning if it retains the spirit of dis-interestedness which ani-
mates the idea of responsibility for the other man" (Levinas 1994, 99).

The detachment with which judgment is to be made has been understood by the Stoics, formulated with some rigidity by Kant, and even-

tually revisited by contemporary thinkers, like Levinas and Lyotard.
For his part, Lyotard quotes Buber to emphasize the inspiration of the
Jewish tradition of ethical behavior, one that dates back to the Ten
Commandments but that is infused with psychosocial wisdom of ex-
perience and turmoil:

> When Jaacob Yitzchak of Lubin conceded to Yeshaya that "when we
> seek to effect nothing, then and then only we may not be wholly with-
> out power" [quoted from Martin Buber], he circumscribed the stakes of
> the genre of ethics: its success (justice) would be the perfect disinterested-
> ness of the ego, the relinquishing of the will. (Lyotard 1988, 129)

Incidentally, Lyotard's echo of Levinas's concern with ethics and justice reverberates in the work of Jacques Derrida, who concedes that
everything (textual or scientific, in his sense) is always inevitably political—regardless of detached judgments and the rules of the law—and
therefore has ethical stakes and consequences (Derrida 1988, 136).

Perhaps it is too much to ask one to follow Levinas's prescription to
take responsibility for the Other, whether one knows that Other or
not; perhaps a modification is in order; perhaps a medium of exchange must be established to make the theoretical principle operable.
Yet despite the exaggerated demand of his ethical system, Levinas
widens the ethical context to such an extent that he overcomes some
of the weaknesses of both Kant's and Lyotard's systems. Instead of a
Kantian abstraction that assumes individual actors without accounting for their individuality, and instead of a Lyotardian principle of de-
liberate demarcation of particular contexts without a demand for the
assumption of responsibility, Levinas gives us a firm directive: Go beyond yourself and reach to another, be responsible for someone other
than yourself! Though this proposal may seem silly at first, it has
much to recommend it: On the one hand, one's extension beyond one-
self enhances a sense of responsibility in one's own life; and on the
other hand, it ensures a sense of community because everyone is constantly thinking about others. In a sense, then, he tries to integrate
moral agency into a regulative ideal.

Levinas turns out to be a promoter of human agency and of the
communal context within which it can be tested and thrive. The French
postmodern trend of the late twentieth century turns out to be much
more sympathetic to human follies and aspirations, their dreams and
the tools without which they cannot fulfill them, than more "progres-

sive" governmental agencies and bureaucrats. There is, then, a possibility of moving from the anxiety of the existential discourse to a discourse about people and their concerns, a discourse about society and civilization, a discourse whose open-endedness does not deny rules. If rules are based on judgments, and if judgments are human creations and as such are prone to be mistaken, then they can still be revised, and their revision indicates the limits of morally acceptable behavior.

The changes in technoscientific discourse, the one I examine here in relation to Auschwitz and Hiroshima, came about because of public outcry. What outraged the public was the breaking of the trust it has invested in the technoscientific community, and its attacks on that community were charged with moral terms and ethical principles. Shortly after the atrocities at concentration camps were revealed, fingers were pointed not only at Hitler's henchmen but also at the technoscientific infrastructure that willingly supported them. Likewise, the horror of Japanese civilians dying from nuclear weapons prompted a public debate about the ethical dimension of technoscience and, as we have seen above, confessions by some members of the technoscientific community. In order to insist that the ethics of technoscience is not a topic limited to past mistakes and the formulation of new rules of conduct and in order to approach the concerns of the public and the technoscientific community alike in terms of mutual expectations (of control, health, and happiness), I suggest we also consider the material conditions within which the technoscientific community operates.

The Economics of Technoscience

As we saw in Chapter 1, the notion of the "invisible hand" accompanies our myths about modernity and about human progress. Capitalism in its continued transformation—post-Fordism, postindustrialism, even postcapitalism—determines the conditions under which public policy and individual decision-making processes take place. To deny the material power and authority exerted by capitalism is to ignore institutional arrangements and organizations and the way they affect their memberships. To ignore these arrangements and their evolution is to deny certain social realities that are definitely within our control. And to deny human control is to avoid responsibility. At this juncture, I wish to recall the *Challenger* incident because it can concretely illustrate some of the more abstract points made here.

On January 28, 1986, after many secret and public launches of satellites into space, a teacher from New Hampshire named Christa McAuliffe was allowed to accompany one of the missions. Her inclusion heralded a new era for space exploration, for it proved to the public that the fruits of technoscience would not be limited to military uses. One may speculate that, in an age of so-called fiscal responsibility, when government expenses are being routinely cut (at least rhetorically if not in fact), a bit of publicity and goodwill was indeed deliberately extended toward the public so as to ensure increased funding for the space program. This speculation is interesting because it suggests that a segment of the technoscientific community has taken an active role in publicizing its wares and its merits to potential customers, the public.

Whereas it is common to speak of the military-industrial complex, the university-industrial complex, or any additional combinations of elite communities that mutually benefit from each other's work and thereby can profit in the marketplace, here we are speaking of a direct appeal to the public. But this appeal, as it turned out in the case of the *Challenger*, backfired for one unfortunate reason: The shuttle exploded just after it took off, with millions of television viewers following its flight as the explosion happened. The tragedy of Ms. McAuliffe's death was unlike that of an Air Force pilot or NASA scientist, for unlike a school teacher, they are expected to risk their lives for their country. Diane Vaughan's account of the disaster is fascinating in two senses: On the one hand, she provides a detailed report of the workings of this technoscientific community so as to explain its behavior, and on the other, she uses terms reminiscent of those used by Arendt. For example, Vaughan says:

> The cause of disaster was a mistake embedded in the banality of organizational life and facilitated by an environment of scarcity and competition, and unprecedented, uncertain technology, incrementalism, patterns of information, routinization, organizational and interorganizational structures, and a complex culture. (Vaughan 1996, xiv)

Arendt's "banality of evil" in the case of Nazi Germany is replaced with the banality of organizational life and the evil it may produce in democratic institutions in peacetime, but they sound and feel strangely alike. What is banal turns out to be fatal in both cases; and what is fatal is understood to be unacceptable but without recourse, as if no one is blameworthy, as if responsibility cannot be ascertained. I do not wish to claim that there is a moral equivalence between the two cases, between Auschwitz and the *Challenger* explosion, but what is the same in both cases is the public perception that in the aftermath of disasters one can easily distinguish between a technoscientific and an ethical discourse or that one can insulate the one from the other. Arendt and Vaughan, in their respective ways, alert us to the unintended consequences of the age of modernity, the age of technological progress and the bureaucratization of everyday life. As Vaughan says: "The practical lessons from the *Challenger* accident warn us about the hazards of living in this technological age" (Vaughan 1996, xv).

Vaughan claims that instead of searching for and finding individual causal explanations concerning the failures and mistakes that led to the *Challenger* disaster, one may more fruitfully consider the following scenario: The rules of the organization were followed, but they were set within a technological culture that tolerated deviance for the sake of efficiency, that encouraged expediency to accommodate financial and political constraints. Summarizing the Presidential Commission's Report of the investigation of the disaster, she says: "In volume 1, however, we repeatedly read that 'NASA' did this or 'NASA' did that—language that simultaneously blames everyone and no one, obscuring who in the huge bureaucracy was actually doing what" (Vaughan 1996, 70). She complains that what gets obscured is the context of cut budgets and political visibility within which middle management has to make decisions that end up less than optimal, at times even dangerous and disastrous.

To speak of economics in the midst of ethics, to bring up financial considerations within the context of technoscience, is to appreciate the fluidity of one's culture and the way humans interact with each other. In the aftermath of disasters, one feels more comfortable asking normative questions: Should we have cut the funding of the space program and thereby endangered the lives of its participants? Should we send civilians to space? Should the public be more involved in the routines of technoscientific projects? Should every member of the technoscientific community be personally accountable and therefore responsible for every decision made within a specified segment of that community?

If we wish, as a society, to respond positively even to some of these questions, that is, if we agree to infuse the technoscientific discourse with some ethics, then we must contend with other questions. The second set of questions is concerned with the practical implications of such a commitment: How would we achieve a heightened sense of accountability and responsibility? Who would monitor whom? Who would pay for additional layers of bureaucracies? If new reporting requirements were to be added, would they in and of themselves in fact accomplish what is expected? Even if we were able to answer positively to some of these questions, still we would have to ask: How would we pay for all of this? In brief, even when we know who pays for the research and development of technoscience, it is unclear whether the same entities would pay for maintaining its moral fiber, its adherence to ethical norms.

The realization that (post)capitalism still dominates our culture cannot be avoided. Both German efficiency in the use of technoscience (from railroads to gases) and U.S. efficiency in the development of nuclear weaponry proved to be lethal. Ethics was relegated to intellectuals and artists, to those concerned with evil as such and not with a particular evil enemy. The fact that intellectuals and artists are prone to question the definition of evil and the particular assignment we make of it in the case of wars (domestic or foreign) separates them from the direct engagement into which technoscientists are invited. It is not only that the latter can and do produce weapons but that they are discouraged from open reflection on the very projects they undertake. Serres taps into this sense of evil in our culture when he says:

> *The problem of evil underlies the power we derive from our various means of addressing it* . . . The future will force [scientific] experts to come quickly to the humanities and to humanity . . . As a result, what is philosophy? The irrepressible witness of universal misfortune before an absolute knowledge that, without this *instruction* (in the multiple senses of origin, pedagogy, and law), would be the equivalent of irresponsible ignorance, whose naivete would reconstruct a new world without forgiveness. (Serres 1995, 182–183)

It would be convenient to blame industry for luring technoscientists into projects that may claim the lives of innocent people; we could blame greed—or human nature—for technoscientists' propensity to overlook moral dilemmas when generous compensations are at hand; we could blame politicians and military leaders for their use of technoscience; and finally, we could blame the public for supporting technoscience for its benefits without assuming the responsibility for undesired effects. Yet in any of these cases, we would be shifting the blame and the responsibility from one party to another, avoiding any deliberation of the confluence of technoscience and ethics. If we agree that this situation is intolerable, then we should change it. Change would require two steps: First, we should involve the public in all technoscientific projects throughout their duration—from inception to research, from development to implementation; and second, if we cannot abolish (post)capitalism (as I doubt we can), we should add ethics in general and some specific principles in particular to the budgets and equations that guide technoscience and the operations of the technoscientific community.

At the end of Chapter 5, I suggested ten behaviors that I think may prevent individual members of the technoscientific community from avoiding a sense of personal accountability and responsibility for the work they do in large groups. In the face of Auschwitz and Hiroshima and in the face of the postmodern age and its insistence on ambiguity, anxiety, and anguish, to ask for codes of ethics is not a luxury but a necessity. But in order to implement even some of these, that is, in order for individuals to adhere to these principles (regardless of one's view of human nature), it may be helpful to provide sufficient funds as incentives to behave ethically or as protective devices that insulate those individuals from suffering because they behaved ethically.

For example, whistle-blowers should be funded while their cases are examined, so that the possibility of being dismissed from their jobs should not frighten them from admitting to immoral corporate or institutional behavior. Likewise, those opposed to projects on moral grounds should have available funds with which to launch alternative projects. And finally, intellectuals and artists should be funded to be part of the technoscientific community so as to encourage the expression of ambiguity, anxiety, and anguish throughout society. Any of these suggestions would render technoscience more expensive in one sense, but no expense is too much to preserve the dignity of humanity into a new century and new millennium.

7

Cultural Changes: Agenda Setting

In order to ascribe responsibility to members of the technoscientific community or to particular groups of leaders (political, military, or academic) or to the (voting and nonvoting) public at large, it may be helpful to rethink the notion of responsibility in its legal setting. If we agree that the legal code follows social conventions and moral norms, then the law could serve as a convenient expression from which to read and ascertain cultural convictions and commitments. Yet for the law to ascribe responsibility, there needs to be a clarification of one's liability in terms of intent. As Frederick Kempin says when discussing torts, there are three different theories according to which courts adjudicate responsibility.

The simplest and most straightforward theory focuses on "the idea of intention. If we intend to injure someone we should be liable to the person injured." One's intention determines one's responsibility. "The second is the idea of fault. A person, under this theory, should not be liable for the consequences of his acts unless he was, in some meaning of the term, at fault." This theory is more complex because the concept of fault can be broadly or narrowly defined and as such is prone to highlight ambiguities. "The third theory is one of absolute liability." This is the most interesting of the theories in the present context because it deflects the ambiguities that may arise from the notions of intent and fault and focuses on the inception of actions regardless of their original intent and eventual fault. "If one starts a course of action which ultimately results in another person's injury he should be liable despite the lack of intention to injure and despite the fact that he was not at fault in any recognizable way" (Kempin 1973, 164).

If one were to follow the third theory described by Kempin, then any technoscientist, American or German, would be found responsible for the mass destruction of human lives at Auschwitz and Hiroshima. The indictment would be categorical in Kant's sense, and no one could escape the wrath of survivors or the judgment of history. (Incidentally, this view comes closer to the view of Dwight Macdonald, who assigns responsibility to scientists, as opposed to the view of Michael Walzer, who blames political leaders, as discussed in Chapter 5.) Just because individual technoscientists could not envision the uses made of their inventions does not mean that they are not responsible for their work in some indirect way, especially if they were funded along the way by those who eventually used their work. This is not to say that there is moral equivalence between Hitler's orders to carry out the Final Solution and the work on nuclear technology of Werner Heisenberg and his colleagues. Perhaps a more apt example of technoscience at the service of mass destruction is the example of the implementation of Zyklon B by researchers who went outside their laboratories to concentration camps, as discussed in Chapter 1.

Yet Kempin's third theory intimates that to exonerate Heisenberg from any responsibility whatsoever for Hiroshima because he neither intended to murder Japanese nor was at fault in some definite way is granting him too much. Heisenberg is responsible, according to this theory, in some legal sense rather than in the moral sense that Levinas develops which sees every human being as ultimately responsible for every other human. But if we found earlier that Levinas's expectations were set too high and were therefore unrealistic, why bring up his ideas once again? Is it because there is a logical equivalence between Levinas's line of reasoning about morality and Kempin's third theory about legality?

Though the legal theory that assigns responsibility may go too far for our contemporary sensibilities (in a fashion similar to the way Levinas's principle does), it may turn out to be a Kantian heuristic device that we desperately need at this juncture of the development of our culture. Perhaps this theory, however demanding, can serve as a starting point or a yardstick against which to measure our conduct (as I outlined at the end of Chapter 5) and our policy-making processes.

Let us say that everyone is always responsible for the actions that will be carried out by others using her or his innocent ideas and research. Would that not foster a more cautious technoscientific community? Would that not enhance a stronger sense of responsibility in everyone? Would that not encourage reciprocity among all in daily life? Could that not lead one to think of one's moral and social agency and establish a sense of community?

To incorporate into one's professional work questions of responsibility (with an appreciation of historical contexts and policy implementation) means to highlight one's self-awareness of a responsible approach to the community. Within this context, it seems to me that postmodernism may enhance the responsibility of intellectuals in general (in their discursive practices) and technoscientists in particular (in their professional practices) by being critical, self-reflective, and skeptically engaged, all at once. This concluding chapter recalls the quest for ordering a chaotic universe within a Stoic model of skeptical acceptance of the human condition. If de Beauvoir is right in declaring that the human condition enhances rather than retards our proactive engagement with each other and with moral concerns, then there may be a better future in store for us. This may be true not despite but because of the deep anguish we feel in the face of technoscience in the postmodern age. If we reject her plea, then the next century and the next millennium will turn out to be more horrible than anything we have experienced so far.

Back to Oppenheimer

Perhaps there is a bit of drama in this closing chapter, a drama worth the name. For what else would one call our situation, a situation in which our survival is at stake? Perhaps those of us who read books in the comfort of our home may feel that pollution and other environmental disasters are far-fetched and far-off from the immediacy of our existence. Only some of us recognize the concerns of de Beauvoir, Lyotard, and Levinas. Yet we would have to admit that the next time gas chambers are built and cattle trains carry humans to their execution or nuclear bombs are dropped on urban centers, we may be the targets, the innocent victims. With the progress of technoscience, the trains will be quicker than they were during World War II, the gas chambers more efficient, the nuclear bombs more lethal, and the targets more widely spread. Moreover, instead of having a world war to spur these horrible events (and thus warn us of our vulnerability), we may find ourselves in the middle of one of those science-fiction novels or movies in which a group of renegades decides, for whatever reason, to torture and kill, to execute and annihilate, everyone in their way.

Modern technoscience makes our self-destruction readily accessible and easily attainable, and postmodern ethics or the responsibility of every single member of the technoscientific community may be our only hope of averting this potentiality. Perhaps we should surround ourselves with a culture of abundance instead of scarcity and thereby eliminate many of the reasons used to justify greed and hatred, thoughtless competition and avarice, in short, class warfare. Perhaps if we had the comfort and balance of meaningful work and moderate leisure, we would all become philosophically minded enough to survive another century, even a millennium. Finally, perhaps if we were all granted the role of cultural critics, regardless of our job description and social stratification, we could alert each other about incidents of injustice and danger for the benefit of society as a whole. Instead of relegating the responsibility of expressing ambiguity, anxiety, and anguish to poets, philosophers, and artists, it is in society's best interest to induce all of its members to express these concerns and by cultivating and praising these expressions of concern, to lead all members of society to realize how important these expressions are for everyone's survival.

In one of his speeches a decade after the end of World War II, Oppenheimer suggests that when speaking of the future of our civilization, we ought to see the parallel situations (and therefore the parallel roles) in which artists and scientists find themselves. Oppenheimer bolsters my own claim that if our culture were to allow the blurring of the roles and expressions of intellectuals, artists, and technoscientists, we would enhance a sense of community, encourage fraternity in Levinas's sense, and include more members of society in a process of reflection and critique. If we were to do all of this, then, to the extent that we all share in the process of making decisions about our life, about our survival, the notion of responsibility would gain new respect and prominence in our midst.

In order to appreciate Oppenheimer's insight, let me quote his words deliberately, a piece at a time. He begins with a general statement about the condition of humanity after World War II: "This balance, this perpetual, precarious, impossible balance between the infinitely open and the intimate, this time—our twentieth century—has been long in coming; but it has come." Here Oppenheimer sounds as if he belonged to the existentialist movement, as if he were reflecting on the human condition. His reference includes Kant's matrix of space and time, as he speaks about the historical period and our relation to space. The horizon remains open and endless, but the immediacy of our surrounding confronts us daily.

"It is, I think, for us and our children, our only way. This is for all men." The way in which he generalizes from his own perception to the entire population, cross-generationally, enforces the philosophical bent of his speech. He speaks, indeed, about the human condition, its reality, immediacy, and eternity.

But after having included the entire human race in his reflection, Oppenheimer pays close attention to two subgroups: artists and scientists. "For the artist and for the scientist there is a special problem and a special hope, for in their extraordinary different ways, in their lives that have increasingly divergent character, there is still a sensed bond, a sensed analogy." Admitting that there is a divergence between these two subgroups, Oppenheimer nonetheless believes that they are similarly plagued by certain concerns and also that they have it in them to tackle the human condition and its problems in a hopeful manner. Once again, in words that echo Albert Camus (1991), Oppenheimer inserts the concept of hope into a dire description of the problems of humanity.

"Both the man of science and the man of art live always at the edge of mystery, surrounded by it; both always, as the measure of their creation, have had to do with the harmonization of what is new with what is familiar, with the balance between novelty and synthesis, with the struggle to make partial order in total chaos." This statement reiterates earlier comments made by the Stoics and by Gellner in regards to the attempt to order a chaotic universe. Oppenheimer alludes to mystery almost in the way Gellner speaks of magic, and he proceeds to speak of creativity and harmony as if artist and scientist alike wonder beyond the confines of everyday life.

But Oppenheimer does not romanticize the notion of creativity as if it were the lonely work of geniuses; instead, he gently moves individuals into a community: "They can, in their work and in their lives, help themselves, help one another, and help all men. They can make the paths that connect the villages of arts and sciences with each other and with the world at large the multiple, varied, precious bonds of a true and world-wide community" (Oppenheimer 1955, 145–146). Far beyond the technoscientific community or the community of artists, there is "a true and world-wide community" to which we all belong. Having witnessed the destruction of people oceans away from the American soil, Oppenheimer seems painfully aware of the direct connection among all members of this planet that we have imposed on ourselves in the postmodern era.

Perhaps Oppenheimer bemoans our condition because he realized after World War II that we no longer had the choice of belonging to a community or not; perhaps he realized that an American team of technoscientists at Los Alamos determined the fate of Japanese citizens, just as a team of German technoscientists determined the fate of Jews all over Europe. Once the postmodern technoscientific world was put in place, there was no way of containing its power and reach. Having wrestled with the gods in order to have greater control over nature by the use of fire and tools, we find ourselves having to share the gods' responsibility for the survival of this universe.

Back to Objectivity and Relativism

To claim that we—the general public, the technoscientific community, the community of intellectuals and political and military leaders—are responsible for the survival of the universe requires some agreement on a set of principles or ideas according to which to measure and enforce our responsibility. It would be horrifying if one were to justify the murder of innocent citizens because someone decided to kill them; it would be likewise horrifying if one were unable to provide a set of criteria according to which frivolous justification of immoral conduct could be ascertained. What, then, is required at the end of this century? According to Gellner:

> What we desperately need is precisely a morality beyond culture, and knowledge beyond morality *and* culture . . . I am absolutely certain that we do indeed possess knowledge beyond both culture and morality. This, as it happens, is both our fortune and our disaster. (Gellner 1992, 54)

Gellner means to imply not that there is anything wrong about the acquisition of knowledge but that the way in which we have approached and fed our obsession with knowledge acquisition has been detrimental to our attitude toward culture and morality. In saying this, he echoes Rousseau's famous admonition of Enlightenment ideals (1964). Rousseau warned against losing the moral courage and sensitivity, the pity and compassion, that characterized the "noble savage." He warned that too much clever learning would turn the noble savage into an immoral educated citizen who could manipulate knowledge against other humans, as opposed to on their behalf.

Gellner is as worried as Rousseau that knowledge has replaced morality, and he therefore recommends that we eschew current trends and revert to a form of "rationalist fundamentalism":

> The mild rationalist fundamentalism which is being commended does not attempt, as the Enlightenment did, to offer a rival counter-model to its religious predecessor. It is fundamentalist only in connection with the form of knowledge, and perhaps in the form of morality, insisting on symmetry of treatment for all . . . Otherwise, on all points of detail and content, it compromises. (Gellner 1992, 94)

So the fundamentalism is a Kantian regulative ideal insofar as whatever principle is agreed upon should apply to everyone equally (Kant's sense of symmetry); and the rationalism is a means by which claims and judgments could be critically scrutinized and openly reasoned. These basic requirements are juxtaposed against the views of what Gellner calls relativists:

> The relativists, in whatever guise—the "postmodernists" are but an extravagant, undisciplined and transient mode of this attitude—seem to me to offer an accurate account of how we do, and probably how we should, order our gastronomy (at any rate on any one evening), our wallpaper, and even, for lack of a better alternative, our daily self-image . . . Their insights apply to the decorative rather than the real structural and functional aspects of our life. (Gellner 1992, 95)

And here Gellner distinguishes, quite arbitrarily, between structure and content, falling prey to the zeal and oversight of the structuralists and functionalists (sociologists and anthropologists alike). Whatever "our daily self-image" may turn out to be, it may be an important ingredient in how we structure our environment and our relations with other humans; and if there is an "accurate account" of how we "order" our material needs (in the double sense of organizing and consuming), then these so-called relativists deserve more praise than they are rendered.

But Gellner is quick to take away a compliment he has almost given to the relativists and turn it into a biting criticism:

> To the relativists, one can only say—you provide an excellent account
> of the manner in which we choose our menu or our wallpaper. As an ac-
> count of the realities of our world and a guide to conduct, your position
> is laughable. Possibly you may be doing good by encouraging political
> compromise. If the ambiguities of your formulations and attitude help
> ease the situation, and bring forth a compromise between the believers
> and the others, or between rival believers, you may still be performing
> a public service. (Gellner 1992, 94–96)

So, the relativists do perform a public service, as he says. They help
bring about a "compromise," or they help translate between competing discourses. Would compromise lead members of the technoscientific community to assume a greater sense of responsibility? One could
argue that there is no room for compromise on the evil of the Holocaust or the atomic bomb. And here one would expect to find, dare I
say, an objective criterion by which to judge the mass murder of innocent people an immoral act!

> Within the whole tradition, it is incidentally possible to discern two dis-
> tinct arguments . . . One of them is that the pursuit of objectivity is re-
> ally spurious, and a form of domination: the observer insulates the ob-
> jects and sits in judgement on it [sic] . . . But there is also the argument
> that the world has become more complex, and that the separation of
> roles is no longer possible (but was once practicable). It is true that the
> world has become more tangled and unstable; but this, to my mind,
> shows only that objectivity is harder, not that it is inherently misguided
> and must be replaced by stylistic chaos and pastiche. If we note that the
> world has changed, we would seem to be in possession of some objec-
> tive information about it after all. (Gellner 1992, 41)

Gellner puts it simply: If one is capable of judging anything, then
one is in fact engaged in an evaluation against some standard, some
yardstick. If one concedes any sense of measurement, then one is indeed admitting to using a foundation, a matrix that is prefigured, or
conventionally agreed upon. And once this is done, then the notion of
objectivity is no longer as abhorrent as it may have sounded earlier.
Perhaps what Gellner speaks of is not so much the human condition
as human conduct, the way people interact and respond to each other,
the way they create communities that survive over periods of time. Perhaps Gellner the anthropologist overrides Gellner the moral philosopher and thus anchors his insights historically and socially in a manner different from the way philosophers, artists, and technoscientists
anchor theirs.

Conclusion

The focus here has been on members of the technoscientific community during World War II, whether American or German, and not on their military and political leaders. It seems that the predicament facing members of this community in their institutional setting extends to the rest of society, thereby exasperating a bad situation rather than diffusing its dangers. Surveying this situation in light of Auschwitz and Hiroshima helps sharpen our vision and our sensibility.

The predicament facing technoscientists is the scientific ethos (as described by Merton). On the one hand, technoscience rests on ambiguity and needs it for its continued research and critical self-evaluation; it is the motor that propels technoscience forward to another set of discoveries and paradigm shifts. On the other hand, ambiguity is understood to be an obstacle in the quest for order. Technoscientists are expected to reduce ambiguity for the rest of their culture, to have clear answers to difficult questions, yet they must admit the limitations of their answers and thereby reintroduce ambiguity into an ordered model of the universe. The frustration of technoscientists parallels that of the public and its representatives.

Until our culture includes some level of built-in ambiguity in its self-image, technoscientists will have to downplay the inherent ambiguity in the human condition. Whether they could be served by the narratives of writers, poets, artists, and intellectuals remains an open question. Perhaps they should take as their model Oppenheimer's view of the parallel role of scientists and artists; perhaps they should refuse to play a role that they cannot play; perhaps they should insist on having the license to express their anxiety about what they know, what they do not know, and what they may find out in the future.

Technoscientists, then, differ little from their intellectual and artistic counterparts in their deeply felt anguish concerning their ability to preserve and save a culture whose future is unpredictable. When their anguish comes through we should listen to them and empathize with them, not banish them to internal professional and social exile. It is difficult to guess what would have happened had we appreciated this feature of technoscientists' existence prior to and during World War II: Would they have had a different outlook on their work and therefore not produced the tools of mass destruction? Yet once we appreciate the anguish of technoscientists, it seems reasonable to ask a question like this, for we would have trained them and ourselves to think about such questions.

Just as I claimed earlier that there is no moral equivalence between the work of German and American technoscientists during World War II, I wish to make clear that I do not find all subgroups of the techno-scientific community to have identical concerns and views about their work. Some members may feel obliged to serve political agendas, while others may remain steadfastly independent thinkers; some may refuse corporate funding, while others will enthusiastically pursue profit-sharing deals. I also want to distinguish between the roles undertaken by some German scientists in the development of Zyklon B for the de-liberate use in gas chambers and by those German doctors who fur-thered the ideals of racial hygiene and the roles undertaken by the sci-entists at Los Alamos. This is not to say that one group is completely free of any responsibility, while the other is not. Instead, I wish to cre-ate an environment in which we can discuss the responsibility of mem-bers of the technoscientific community in a sophisticated manner: Under what conditions must one refuse to contribute to technoscience? What safeguards does the culture provide? What rules are set up by the leadership of the technoscientific community? What power rela-tions must obtain to ensure avoidance of catastrophic results?

Since I find these questions important not only for rethinking the past but also as guides for the future, I believe philosophers and intel-lectuals must be at the forefront of suggesting the means by which to answer them. I introduced early in this book the notion of philosopher as translator and explained the need to translate between diverse dis-courses. In some ways, I have tried to do this here, quoting from archival sources, diaries, speeches, learned treatises, and plays. I have violated some of the scholarly rules and strayed to fields outside tradi-tional philosophy; I brought together elements that I thought could shed light on each other in new ways. I followed, to some extent, the advice of Serres:

> The philosopher is not a judge; if he is a judge, a critic, he never pro-duces anything, he only kills. No. Trying to think, trying to produce, presupposes the taking of risks, the living of one's life, precisely, in the surge outside of the classing of the encyclopedias. Let us then introduce the concept of chaos. (Serres 1995, 98)

No, I am not a judge nor a legislator in Bauman's sense; instead, I remain an interpreter and translator who tries to walk across a frightening abyss so as not to fall into it and be lost. What can be accomplished by this effort? What can postmodern technoscience achieve for the next millennium? Perhaps all that can be accomplished is a shift in the focus of attention: Instead of asking the unanswerable question "Is this technology good or bad, benign or dangerous?" we should ask "Who is responsible for the production and distribution of this technology?" No technoscientific feat can ever be fully prefigured, and therefore it makes more sense to approach it from the perspective of responsibility. This is not to say that we have resolved all questions about how to define and formulate the notion and expression of responsibility.

Our technoscientific condition should be openly recognized and scrutinized, just as should our human condition, in de Beauvoir's sense. Likewise, we should use the anguish that accompanies the technoscientific condition in the age of postmodernity as an inducement to choose wisely, responsibly, between alternative options. This is a recommendation we cannot afford to ignore, since it is not for the improvement of our cultural predicaments but for our survival.

References

Adorno, Theodor. 1973. "After Auschwitz." In *Negative Dialectics* [1966], 361–365. Trans. E. B. Ashton. New York: Continuum.

Annas, George J., and Michael Grodin. 1992. *The Nazi Doctors and the Nuremberg Code: Human Rights in Human Experimentation.* New York and Oxford: Oxford University Press.

Arendt, Hannah. 1963. *Eichmann in Jerusalem: A Report on the Banality of Evil.* New York: Viking Press.

Bacon, Francis. 1985. *The New Organon* [1620]. New York: Macmillan.

Bauman, Zygmunt. 1987. *Legislators and Interpreters: On Modernity, Post-Modernity, and Intellectuals.* Ithaca, N.Y.: Cornell University Press.

———. 1991a. *Modernity and the Holocaust* [1989]. Ithaca, N.Y.: Cornell University Press.

———. 1991b. *Modernity and Ambivalence.* Ithaca, N.Y.: Cornell University Press.

———. 1992. *Intimations of Postmodernity.* London and New York: Routledge.

Baumol, William J., John C. Panzar, and Robert D. Willig. 1982. *Contestable Markets and the Theory of Industry Structure.* New York: Harcourt Brace Jovanovich.

Bentley, Eric. 1966. Introduction to Bertolt Brecht, *Galileo* [1940/1952]. Trans. Charles Laughton. New York: Grove Press, 1966.

Berkson, William, and John Wettersten. 1984. *Learning from Error: Karl Popper's Psychology of Learning.* La Salle, Ill.: Open Court.

Bernstein, Jeremy. 1996. *Hitler's Uranium Club: The Secret Recordings at Farm Hall.* Woodbury, N.Y.: American Institute of Physics.

Bourdieu, Pierre, and Loic J. D. Wacquant. 1992. *An Invitation to Reflexive Sociology.* Chicago: University of Chicago Press.

Brecht, Bertolt. 1966. *Galileo* [1940/1952]. Trans. Charles Laughton. New York: Grove Press.

Buber, Martin. 1970. *I and Thou* [1922]. Trans. Walter Kaufmann. New York: Charles Scribner's Sons.

Butterfield, Herbert. 1965. *The Origins of Modern Science, 1300–1800* [1957]. New York: Free Press.

Camus, Albert. 1991. *The Myth of Sisyphus and Other Essays* [1942]. Trans. Justin O'Brien. New York: Vintage.

Cassidy, David C. 1992. *Uncertainty: The Life and Science of Werner Heisenberg*. New York: W. H. Freeman.

Dahl, Robert. 1986. "Power as the Control of Behavior" [1968]. In *Power*, ed. Steven Lukes, 37–58. Oxford: Basil Blackwell.

Davis, Nuel Pharr. 1968. *Lawrence and Oppenheimer*. New York: Simon and Schuster.

Dawidowicz, Lucy S. 1975. *The War against the Jews, 1933–1945*. New York: Bantam Books.

de Beauvoir, Simone. 1991. *The Ethics of Ambiguity* [1948]. Trans. B. Frechtman. New York: Citadel Press.

de Condorcet, A-N. 1979. *Sketch for a Historical Picture of the Progress of the Human Mind* [1795]. Trans. J. Barraclough. Westport, Conn.: Greenwood Press.

Delkeskamp-Hayes, Corinna, and Mary Ann Gardell Cutter, eds. 1993. *Science, Technology, and the Art of Medicine*. Dordrecht and Boston: Kluwer.

Derrida, Jacques. 1988. *Limited Inc* [1977]. Trans. Samuel Weber and Jeffrey Mehlman. Evanston, Ill.: Northwestern University Press.

Dewey, John. 1929. *The Quest for Certainty: A Study of the Relation of Knowledge and Action*. New York: Capricorn Books.

Duhem, Pierre. 1969. *To Save the Phenomena*. Trans. E. Doland and C. Maschler. Chicago: University of Chicago Press.

Durrenmatt, Friedrich. 1964. *The Physicists* [1962]. Trans. James Kirkup. New York: Grove Press.

Eatherly, Claude, and Günther Anders. 1989. *Burning Conscience: The Guilt of Hiroshima* [1961]. New York: Paragon House.

Epictetus. 1983. *The Handbook (The Encheiridion)* [first century A.D.]. Trans. Nicolas White. Indianapolis, Ind.: Hackett.

Fackenheim, Emile. 1970. *God's Presence in History: Jewish Affirmations and Philosophical Reflections*. New York: Harper Torchbooks.

Fein, Helen. 1979. *Accounting for Genocide: National Responses and Jewish Victimization during the Holocaust*. Chicago and London: University of Chicago Press.

Freud, Sigmund. 1989. *Civilization and Its Discontents* [1930]. Trans. James Strachey. New York: Norton.

Gadamer, Hans-Georg. 1989. *Truth and Method* [1960], 2nd rev. ed. Trans. J. Weinsheimer and D. G. Marshall. New York: Crossroad.

Galton, Francis. 1908. *Memories of My Life*. London: Methuen.

Gellner, Ernest. 1974. *Legitimation of Belief*. Cambridge, Eng.: Cambridge University Press.

———. 1992. *Postmodernism, Reason, and Religion*. London and New York: Routledge.

Gilbert, Martin. 1985. *The Holocaust: A History of the Jews of Europe during the Second World War*. New York: Holt.

Goldhagen, Daniel Jonah. 1996. *Hitler's Willing Executioners: Ordinary Germans and the Holocaust*. New York: Knopf.

Groves, Leslie R. 1962. *Now It Can Be Told: The Story of the Manhattan Project*. New York and Evanston: Harper and Row.

Guattari, Félix. 1986. "The Postmodern Dead-End." *Flash Art* 128:40–41.

Habermas, Jürgen. 1979. *Communication and the Evolution of Society* [1976]. Trans. T. McCarthy. Boston: Beacon Press.

Hall, Stuart. 1992. "Cultural Studies and Its Theoretical Legacies." In *Cultural Studies*, ed. L. Grossberg, C. Nelson, and P. Treichler, 277–294. New York: Routledge.

Haraway, Donna. 1992. "Otherworldly Conversations; Terran Topics; Local Terms." *Science as Culture* 3(14):64–98.

Hawkins, David. 1983. "Toward Trinity" [1946]. Part I of *Project Y: The Los Alamos Story*. Series in the History of Modern Physics, 1800–1950. Los Angeles and San Francisco: Tomash.

Hegel, G. W. F. 1967. *Hegel's Philosophy of Right* [1821]. Trans. T. M. Knox. Oxford: Oxford University Press.

Heisenberg, Werner. 1971. *Physics and Beyond*. Trans. A. J. Pomerans. New York: Harper and Row.

Hilberg, Raul. 1973. *The Destruction of the European Jews*. New York: New Viewpoints.

Hoesterey, Ingeborg, ed. 1991. *Zeitgeist in Babel: The Post-Modern Controversy*. Bloomington and Indianapolis: Indiana University Press.

Holton, Gerald. 1996. *Einstein, History, and Other Passions* [1995]. Reading, Mass.: Addison-Wesley.

Irwin, Alan. 1995. *Citizen Science: A Study of People, Expertise, and Sustainable Development*. London and New York: Routledge.

Kant, Immanuel. 1970. "An Answer to the Question: 'What is Enlightenment?'" In *Kant's Political Writings*. Trans. H. B. Nisbet, 41–48. Cambridge, Eng.: Cambridge University Press.

Kempin, Frederick G., Jr. 1973. *Historical Introduction to Anglo-American Law*. St. Paul, Minn.: West.

Kipphardt, Heinar. 1968. *In the Matter of J. Robert Oppenheimer: A Play Freely Adapted on the Basis of the Documents* [1964]. Trans. Ruth Speirs. New York: Hill and Wang.

Kolakowski, Leszek. 1990. *Modernity on Endless Trial*. Chicago and London: University of Chicago Press.

Kuhn, Thomas S. 1970. *The Structure of Scientific Revolutions* [1962]. Chicago: University of Chicago Press.

Latour, Bruno, and S. Woolgar. 1986. *Laboratory Life: The Construction of Scientific Facts* [1979]. Princeton: Princeton University Press.

Levinas, Emmanuel. 1969. *Totality and Infinity: An Essay on Exteriority* [1961]. Trans. Alphonso Lingis. Pittsburgh, Pa.: Duquesne University Press.

———. 1985. *Ethics and Infinity: Conversations with Philippe Nemo* [1982]. Trans. Richard A. Cohen. Pittsburgh, Pa.: Duquesne University Press.

———. 1994. *Beyond the Verse: Talmudic Readings and Lectures* [1982]. Trans. Gary D. Mole. Bloomington and Indianapolis: Indiana University Press.

Lyotard, J.-F. 1982. "New Technologies." In *Political Writings* [1993]. Trans. B. Readings and K. Paul. Minneapolis: University of Minnesota Press.

———. 1984. *The Postmodern Condition: A Report on Knowledge* [1979]. Trans. G. Bennington and B. Massumi. Minneapolis: University of Minnesota Press.

———. 1988. *The Differend: Phrases in Dispute* [1983]. Trans. G. Van Den Abbeele. Minneapolis: University of Minnesota Press.

Lyotard, J.-F., and Jean-Loup Thébaud. 1985. *Just Gaming* [1979]. Trans. W. Godzich. Minneapolis: University of Minnesota Press.

Macdonald, Dwight. 1953. *The Root Is Man: Two Essays in Politics* [1945–1946]. Alhambra, Calif.: Cunningham Press.

Marx, Leo. 1964. *The Machine in the Garden: Technology and the Pastoral Ideal in America*. Oxford: Oxford University Press.

Merton, Robert. 1973. *The Sociology of Science: Theoretical and Empirical Investigations*. Chicago and London: University of Chicago Press.

Mill, John Stuart. 1984. *On Liberty* [1859]. Harmondsworth, Eng.: Penguin Books.

Morone, Joseph G., and Edward J. Woodhouse. 1989. *The Demise of Nuclear Energy? Lessons for Democratic Control of Technology.* New Haven and London: Yale University Press.

Oppenheimer, J. Robert. 1955. *The Open Mind.* New York: Simon and Schuster.

Ormiston, Gayle, and Raphael Sassower. 1989. *Narrative Experiments: The Discursive Authority of Science and Technology.* Minneapolis: University of Minnesota Press.

Orwell, George. 1961. *1984* [1949]. New York: New American Library.

Pickering, Andrew, ed. 1992. *Science as Practice and Culture.* Chicago and London: University of Chicago Press.

Polanyi, Michael. 1966. *The Tacit Dimension.* New York: Doubleday.

Popper, Karl. 1959. *The Logic of Scientific Discovery* [1935]. New York: Harper and Row.

———. 1963. *Conjectures and Refutations: The Growth of Scientific Knowledge.* New York: Harper and Row.

Proctor, Robert. 1988. *Racial Hygiene: Medicine under the Nazis.* Cambridge, Mass., and London: Harvard University Press.

Rosenau, Pauline Marie. 1992. *Post-Modernism and the Social Sciences.* Princeton: Princeton University Press.

Rousseau, Jean-Jacques. 1964. "Discourse on the Origins and Foundations of Inequality." In *The First and Second Discourses,* ed. Roger D. Masters and trans. Judith R. Masters, 99–248. New York: St. Martin's Press.

Russell, Bertrand. 1960. *Authority and the Individual* [1949]. Boston: Beacon Press.

Sartre, Jean-Paul. 1962. *Anti-Semite and Jew* [1946]. Trans. G. J. Becker. New York: Grove Press.

———. 1976. *No Exit* [1945]. In *No Exit and Three Other Plays.* New York: Vintage.

Sassower, Raphael. 1993. *Knowledge without Expertise: On the Status of Scientists.* Albany: State University of New York Press.

———. 1994. "The Politics of Situating Knowledge: An Exercise in Social Epistemology." *Argumentation* 8:185–198.

———. 1995. *Cultural Collisions: Postmodern Technoscience.* New York: Routledge.

Sassower, Raphael, and Michael Grodin. 1987. "Scientific Uncertainty and Medical Responsibility." *Theoretical Medicine* 8:221–234.

Schlick, Moritz. 1985. *General Theory of Knowledge* [1925]. Trans. A. E. Blumberg. La Salle, Ill.: Open Court.

Sellars, Wilfrid, and John Hospers, eds. 1970. *Readings in Ethical Theory,* 2nd ed. Englewood Cliffs, N.J.: Prentice-Hall.

Serres, Michel. 1995. *Genesis* [1982]. Trans. Genevieve James and James Nielson. Ann Arbor: University of Michigan Press.

Smith, Adam. 1937. *An Inquiry into the Nature and Causes of the Wealth of Nations* [1776]. Ed. E. Cannan. New York: Modern Library.

Smith, Alice Kimball, and Charles Weiner, eds. 1980. *Robert Oppenheimer: Letters and Recollections.* Cambridge, Mass., and London: Harvard University Press.

Snow, C. P. 1964. *The Two Cultures and a Second Look.* Cambridge, Eng.: Cambridge University Press.

Society for Social Responsibility in Science (SSRS). 1971. *Newsletter,* no. 4.

Stephan, Paula, and Sharon Levin. 1992. *Striking the Mother Lode in Science: The Importance of Age, Place, and Time.* New York and Oxford: Oxford University Press.

Sterba, James P., ed. 1984. *Morality in Practice.* Belmont, Calif.: Wadsworth.

Strickland, Donald. 1968. *Scientists in Politics: The Atomic Scientists Movement, 1945–46.* West Lafayette, Ind.: Purdue University Studies.

Suleiman, Susan Rubin. 1994. *Risking Who One Is: Encounters with Contemporary Art and Literature*. Cambridge, Mass., and London: Harvard University Press.

Szasz, Ferenc Morton. 1984. *The Day the Sun Rose Twice: The Story of the Trinity Site Nuclear Explosion, July 16, 1945*. Albuquerque: University of New Mexico Press.

Teller, Edward. 1964. *The Reluctant Revolutionary* [1960]. Columbia: University of Missouri Press.

Toynbee, Arnold. 1956. *The Industrial Revolution* [1884]. Boston: Beacon Press.

Truslow, Edith C., and Ralph Carlisle Smith. 1983. "Beyond Trinity" [1947]. Part II of *Project Y: The Los Alamos Story*. Series in the History of Modern Physics, 1800–1950. Los Angeles and San Francisco: Tomash.

Vaughan, Diane. 1996. *The Challenger Launch Decision: Risky Technology, Culture, and Deviance at NASA*. Chicago and London: University of Chicago Press.

Walzer, Michael. 1977. *Just and Unjust Wars: A Moral Argument with Historical Illustrations*. New York: Basic Books.

Weber, Max. 1958. *The Protestant Ethic and the Spirit of Capitalism* [1904–1905]. Trans. Talcott Parsons. New York: Charles Scribner's Sons.

——. 1968. *Economy and Society: An Outline of Interpretive Sociology*, 2 vols. Berkeley, Los Angeles, and London: University of California Press.

Weisskopf, Victor. 1991. *The Joy of Insight: Passions of a Physicist*. New York: Basic Books.

Wiener, Norbert. 1989. *The Human Use of Human Beings* [1950]. London: Free Association Books.

Winner, Langdon. 1977. *Autonomous Technology: Technics-Out-of-Control as a Theme in Political Thought*. Cambridge, Mass., and London: MIT Press.

Wittgenstein, Ludwig. 1958. *Philosophical Investigations*. Trans. G. E. M. Anscombe. New York: Macmillan.

Index

Raphael Sassower is professor of philosophy at the University of Colorado at Colorado Springs. His latest books are *Knowledge without Expertise: On the Status of Scientists* and *Cultural Collisions: Postmodern Technoscience*.